IN GOOD SHAPE

Style in Industrial Products
1900 to 1960

Stephen Bayley

D1335967

THE
DESIGN
COUNCIL

In Good Shape
Style in Industrial Products 1900 to 1960

First edition published in the
United Kingdom 1979 by
Design Council
28 Haymarket, London SW1Y 4SU

Designed by Gill Streater

Phototypesetting by
SIOS Limited
111-115 Salusbury Road, London NW6 6RJ

Printed and bound in the United Kingdom by
The Whitefriars Press Limited

Distributed in the United Kingdom by
Heinemann Educational Books Ltd
22 Bedford Square, London WC1B 3HH

All rights reserved. No part of this publication may
be reproduced, stored in a retrieval system, or
transmitted, in any form or by any means, elec-
tronic, mechanical, photocopying, recording or
otherwise, without the prior permission of the
Design Council.

© Stephen Bayley 1979

British Library CIP Data

In Good Shape
   1. Design, Industrial – History – 20th century –
Sources
   I. Bayley, Stephen II. Design Council
745.2'09'04 TS171

ISBN 0 85072 095 8
ISBN 0 85072 096 6 Pbk

IN GOOD SHAPE

CONTENTS

'Design' from Cesare Ripa's *Iconologia*, 1625

FOREWORD

by Lord Reilly

Stephen Bayley has been good enough to record my succession to the Directorship of the Council of Industrial Design as the final entry on his 1900 to 1960 date chart, a compliment for which I am grateful, but one which I can frankly return by saying that had I had the opportunity to read a book like *In Good Shape* before taking over from Sir Gordon Russell, I might well have become a wiser, less rigid director of that institution.

I say that because, towards the end of my period in Haymarket, I began to find myself out of touch and indeed out of sympathy with some of the more popular and to me more frivolous manifestations of modern design. I suppose the reason was that I belonged to the generation which was brought up on the preaching and practice of the pioneer members of the Design and Industries Association. I believed wholeheartedly, and still do, in fitness for purpose, truth to materials, honest workmanship and all those other sober-sided, functional yardsticks that distinguished the then avant garde from the run-of-the-mill commercial folk, whose pre- and post-war clichés seem latterly to have been regurgitated by various opportunists in the name of popular culture.

The advantage that someone of Stephen Bayley's generation has over someone of mine is that he can view the swings of taste and fashion more dispassionately and, if he happens to be a Stephen Bayley, more analytically. Mr Bayley is that rare bird a historian of modern design – and that even rarer bird a historian born after the 1951 Festival of Britain. He can thus describe most of the industrial design landmarks of the 60 years spanned by *In Good Shape* quite independently of any personal involvement.

His book, to my mind, gains enormously from this detachment, but even more perhaps from the unusual formula he has adopted. It was a brilliant idea to study those 60 formative years through the mouths and eyes of the main participants. It was also a generous idea, since he now shares his wide reading and his intelligent choice of illustrations with all who are interested in twentieth-century industrial design, and particularly with those who may be professionally involved as teachers or lecturers. Indeed, fascinating as his book is to me in my retirement, it would have been of riveting interest and value to me in mid-career. Stephen Bayley's contemporaries are fortunate to have such a clear-sighted guide who can also write with such clarity.

PREFACE

Industrial design is an expression now current in every European language. It is the name given to one of the most significant processes to have developed during the twentieth century. At one end it can be concerned with the engineering design, choice of materials and ergonomic characteristics of, say, a centre lathe or a dentist's chair, and at the other, with the appearance and style of a fountain pen. This book is about one aspect of industrial design: it is concerned in particular with product design, the business of giving consumer goods an individual identity or 'look'. That is not to say that this is the most important part of an industrial designer's work, but it is the one whose character fits the argument of this book.

It has grown out of a sense of outrage which has increased year by year as I look around and see how art is offered to the public. Increasingly lavish books are published and ever more complicated and expensive exhibitions are organised to create and maintain an interpretation of 'art' which has no basis in public taste. The suggestion of significance is given by glamorous production of monographs about painters whose obscurantist works only ever arrest the attention of a minority willing, in any case, to learn. The public as a whole is quite ignorant of an art which is maintained only by critics with an interest in perpetuating an interpretation of it which died at the end of the last century.

Now, while I would reject the idea that mere aggregates of popularity and acceptance justify a claim for attention, it has struck me as absurd that those sophisticated enough to distinguish between, say, a Braque and a Mondrian painting, are entirely unaware of the names of the men who have shaped their lives and their environment and who have created the familiar appearance of the everyday world. At one time in the 1940s or the 1950s, for instance, it would have been possible for some American citizens to spend every day of their lives entirely surrounded by artefacts and machines whose appearance was the responsibility of one man – Raymond Loewy – and whose performance corresponded exactly to criteria which he had defined. Increased competition in the design business which Loewy had done so much to create means that this is now no longer possible. The visually aware know that the world we inhabit has been designed not only by Raymond Loewy, but also by Kenneth Parker (who helped me write this), by Eliot Noyes (whose work has helped me to type it), by Pierre Boulanger (whose engineering skill has helped me deliver the manuscript) and by many others besides, whose powerful visual imaginations and persuasive intellectual skills have added a richness and facility to the circumstances of everyday life.

Designers will always stress that their business is really just a matter of thinking but, of course, it is art as well. The denial by a designer that artistic considerations play any part in his visual decisions is a type of self-deprecation common to all polemicists. Industrial design is the art of the twentieth century. We need a new word to describe what the national and commercial galleries support. I believe firmly that the man who designed the Julius tomb would find that he had more in common with the methods, processes and aims that Raymond Loewy employed while recasting the Gestetner duplicator, than with someone stacking bricks in the name of 'art'. I feel that to design a vacuum cleaner which remains more or less unchanged visually and mechanically for nearly 60 years is a towering intellectual achievement, and that is why I have compiled this book about the men who have made that their business.

In one sense, at least, this is a very traditional book. The argument presented here that machines can be considered as art is not a very new one. While I am not trying to revivify the old Modern Movement idea that *any* machine which works well looks well too, I do believe that the most skilfully styled machines are mute witnesses to my modern interpretation of traditional art. In this context I take art to be a beautiful and original visual solution to a pre-existing problem.

In this selection of pictures is, I believe, art. I do not entirely reject Louis Sullivan's persuasive credo that 'form follows function', as do some who are emotionally unprepared to enjoy modern architecture and design, but I do treat it with historical distance and with caution. The selection of photographs made here is intended to show peculiarly modern examples of the beautiful which, to paraphrase William Morris, are known by their popularity to have proved themselves useful as well.

The 60 years of industrial design that *In Good Shape* covers have accommodated some astonishing reversals. A too easy popularisation of Modern Movement rhetoric prepared the world to accept the highly tendentious principle that the beauty of machines was inevitable, an idea which writers during the 1920s and 1930s espoused with such virulence that the smoke from the fires in their bellies blinded their view. There are so many variables in both the intellectual and visual processes which comprise 'design' that it would be naive to maintain, too strictly, that a machine which works efficiently has followed immutable laws of functional supply and demand which render it beautiful. Similarly, experience has forced us to realise that the Modern Movement's article of faith that there should be 'truth to materials' has had to be reconsidered. With the number of synthetic materials and finishing processes now available this argument is absurd: what sort of truth, you demand, does chromium-plated vinyl beg to express? With the loosening of these strict constructions comes the idea, alien to the hardliners of the 1920s and 1930s, that the shape of machines is a product of the will of the man who designed them, and not a function of inevitability.

Many of the ideas, like Sullivan's, which have influenced designers across this century have been produced by architects. Indeed, many of the machines illustrated in this book were designed by men who had had either training or practice in architecture. One change this century has seen is the removal of architecture from being the mother of the arts to being the stepchild of design. By the time a building became an industrialised process with relatively little space for manoeuvring into free expression, architecture could no longer be considered the legitimate, total expression of the taste of an age. That architecture today enjoys a lower status than design can be seen in the corporate policies of big industry. In the design programmes set up by some of the immense American corporations architecture is just one part of a whole package which includes product and graphic design as part of a co-ordinated campaign. What Americans call 'name' architects, such as Gropius, Mies or Philip Johnson in the last generation, are just contributors to the corporate identity of Pan-Am, Seagram or Lever.

This, however, is tangential to the centre of this book. The texts and the photographs incorporated here are intended to show how ideas about the look of machines have developed in a crucial 60-year period. The dates are chosen with reason: before 1900 there was no such thing as product design; machines used decoration wholly derived from the vocabulary of architecture or, on the other hand, they were simply the direct products of industrial processes or craft traditions. After about 1900 two things happened: the first was that a self-consciousness about the appearance of ever more familiar machines developed; and the second was a flurry of new inventions that reached the consumer market before their novelty had been diminished by a readily understandable architectural disguise. In between then and 1960 an identifiable, evolving visual language about the look of machines was created in all the industrialised nations.

Are any principles apparent in the development described in the photographs? Products have become cleaner, perhaps more organic, but the precise manner in which taste changes is as unknown to us as it was to Burckhardt or Wölfflin. The shape of, say, a motor car with its wheels faired into the bodywork is obvious to a child today, but 40 years ago it eluded one of the greatest architects of the century.

Viewed synoptically, the Modern Movement's insistence on function and on truth were only temporary obsessions. Some of the best industrial design is, judged by the steely point of view of the 1920s, a fraud and a deception. Judged by our standards it is the best art we have. Industrial designers have produced many beautiful, timeless shapes. If this seems a careless claim in a period of shifting values, then just let me reassert that this is a very traditional book.

Stephen Bayley
January 1979

'The automobile manufacturers have made, in the past few years, a greater contribution to the art of comfortable seating than chair builders had made in all preceding history'

Walter Dorwin Teague, 1940

Writing to his wife one day late in the summer of 1926, the painter Lyonel Feininger broke off from a lyrical passage about a landscape he had only recently seen to describe to her the details of the wing construction of a pair of Junkers flying boats which were lying in the local harbour. Feininger's concern was not parody, nor to divert, but it was typical of an attitude to mechanical beauty which engaged many artists at the beginning of this century. Time and again in reading what architects and painters have written we find references to the machines that have inspired their mechanical romanticism, referred to always in tones of hushed reverence. To the architects and painters of the Modern Movement devotion to the machine was an article of faith. Feininger's was just one of these devotions.[1]

Most familiar, perhaps, of all the artistic spirits which stumbled across the startling beauty of the anonymous machinery he began to notice all around him, was the Swiss architect Le Corbusier, who had been introduced to the beauty of aircraft just after the First World War, while working on his magazine, *L'Esprit Nouveau*. Le Corbusier, as a one-time employee of the Voisin factory, knew well enough that aircraft had proved themselves of military value during the war and popular interest in them was running at an enthusiastic level. It was an aviation exhibition in London which first attracted Le Corbusier's attention and soon there followed in *L'Esprit Nouveau* the series 'Eyes which do not see', where the architect cited steamships, motor cars and aircraft as models of the beauty, precision and technique which buildings should imitate. Soon afterwards he was to make his famous mechanical equation that a house, at least so far as he was concerned, should be treated like a machine for living in. By 1935 he was writing 'the aeroplane points an accusing finger' ('l'avion accuse'), declaring aircraft to be the symbols of the new age of which he considered himself to be the chief prophet.

His reach overstretched his grasp. Although professedly at one with the engineers, Le Corbusier did not really understand what was going on about him. It was not so much that he wanted to apply mechanical techniques to his buildings (many of them, in fact, employed a very large proportion of hand finishing), but that his restless, inquisitive and messianic imagination seized upon mechanical imagery to make an aesthetic argument which industrial and commercial developments over which he had no control had, in any case, made inevitable. His enthusiasm for machinery was as arbitrary as a jackdaw's, a characteristic shared by other artists, such as Fernand Léger or Francis Picabia, who also enjoyed using mechanistic motifs in their art. Léger would quote a Bugatti crankshaft to aid his dogmatising, while Picabia, sensing the sinister mystery of machines, might use internal combustion engine valves as sexual metaphors. All three artists were eloquent, witty and ineffectual witnesses of one of the most significant cultural phenomena of this age, the assumption by industry and life of art's power to astonish us. Artists like Léger and Picabia were the first to sense the dominant role which machines were to play in the creation of the values and sensibilities of the twentieth century.

The selection of machines as exemplars by artists was the first expression of a development which has not finished today. Sheet steel, dollars and a retainer from the Board have replaced oil, canvas and a papal commission as the media and conditions of art, although the talent and processes employed have remained the same. One publishing programme, an abortive one, symbolised the crux of this change in emphasis. It was the origination of two volumes by *The Studio* magazine in 1936 and 1937. The first was by an artist, a member of the old guard who regarded machines as a means, not an end. The second was by an individual designer, a man who made machines the art of the twentieth century. Both books were about similar subjects: the first was *Aircraft*, written by Le Corbusier; and the second was *The Locomotive*, by the industrial designer Raymond Loewy.

Other books in this series were planned, but the Second World War prevented their appearance. The war years can be taken as a watershed for the relegation of art in favour of industrial design, but both the developments which were to establish design as the major legitimate concern of those who cared about the appearance of the world occurred in the years before 1939. The first was the creation of national bodies which, while not actually manufacturing goods themselves, concerned themselves with design as a part of the industrial process. This was a European development and took place in Germany and Britain. The bodies were the Deutscher Werkbund in Germany, and the Design and Industries Association in England. The second development was an entirely independent American one, and characteristically enterprising, which involved the creation by private individuals of design studios which sold their services to manufacturing industry.

This introduction sketches the history of these developments, and the attitudes to design which they entailed in both Europe and America. The creation of the Design and Industries Association and, say, the opening of Raymond Loewy's studio in New York were events as characteristic of the design process on both sides of the Atlantic as

Still from Le Ballet Mécanique *directed by Fernand Léger (Cinegate Ltd)*

'Ici, c'est ici Stieglitz, foi et amour' by Francis Picabia, 1915 (Metropolitan Museum of Art, The Alfred Stieglitz Collection, 1949)

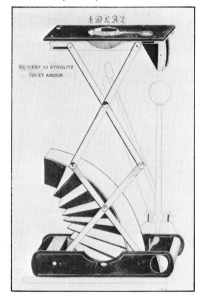

'Fiat 1400' by Giorgio de Chirico, 1950 (Centro Storico Fiat, Turin)

Title page of Aircraft *by Le Corbusier, Studio, 1936*
Title page of The Locomotive *by Raymond Loewy, Studio, 1937 (Photos Design Council)*

Gentleman's Relish and the Breakfast of Champions are characteristic of the respective palates. Both were indicative of the shift of artistic resources from galleries to factories; the change of emphasis from style to technology is another. In building, this could be said to have begun when William Le Baron Jenney started using railroad track in Chicago office buildings and has by no means ended yet. When building plays industry's game, it is not surprising that industrial design with the enormous financial resources which back it begins to replace architecture at the head of the visual hierarchy. The work of artists still acknowledges this. No longer do painters archly cite the machine as an example to us all, and carry on with their traditional solipsistic preoccupations, but today someone like Don Eddy stands back incuriously and lets us wonder at the beauty of the machinery which his paintings so coldly and so clearly describe.

It is almost impossible to disentangle the trails of origin of the design-conscious bodies that emerged in Britain and Germany at the beginning of the century. It was in these highly industrialised countries where the interest in design was greatest and, consequently, where the influence of design theory was porportionately the largest. The ideals of William Morris and the sensible architecture of the English Domestic Revival were a common source which were especially influential in the German-speaking countries where Hermann Muthesius, in books like his *Das Englische Haus* and *Die Englische Baukunst der Gegenwart*, helped spread the ideas and images which had contributed to recent British design. This enthusiasm for the English example led, via the creation of institutions such as the Deutscher Werkbund and the Bauhaus, to Germany's temporarily unrivalled position of influence in the visual arts — an idea made familiar by Pevsner's shrewd and persuasive arguments in books like *Pioneers of Modern Design* and *Academies of Art Past and Present* which are still hard to challenge today.

It was the Deutscher Werkbund which was perhaps the single most significant institutional creation for the progress of industrial design in Europe before the Second World War. The productions of rarefied avant-garde groups like the Wiener Werkstätte, which historians have sometimes tried to emphasise, were irrelevant in the history of the development of modern European industrial design. This was rather as one might expect from an organisation modelled on the example of C R Ashbee's London Guild of Handicrafts. Even if the Werkstätte's members – including of course its auspicious founder, Josef Hoffmann – were able to design in an aggressively modern idiom, their production was by hand and confined to inert products with no moving mechanical parts.

It was the Deutscher Werkbund, founded in Munich in 1906, which established the pattern for design in this century. Josef Hoffmann actually designed the Austrian Pavilion at the Werkbund exhibition in Cologne in 1914, but the organisation's whole tone was intended generally to exclude handicraft and its professed aims were to marry art to industry, which it proposed to do by an enlightened programme of education. The Werkbund aimed to improve German industrial design in the period when the full potential of manufacturing industry was first being felt. It was a novel experience, and writers associated with the Werkbund struggled in heroic articles to establish the rationale of the first conscious genre of industrial design.[2]

England was the only other European country which compared to Germany in terms of industrial might, and an identical body to the Werkbund was created there with the same ends in view. This was the Design and Industries Association (DIA), set up in 1915 in London, with a branch in Manchester and local offices in Bristol, Birmingham and Nottingham. At this time, during the First World War, patriotic considerations did not allow the members of the DIA to expose their ideological debt to the example of Germany, but the common ground was clear. References, not always acknowledged, to ideas first aired by William Morris and Louis Sullivan, together with righteous clamourings about the visual horror of marbled wallpaper and biscuit tins designed to look like a set of golf-clubs or six volumes of Shakespeare, were the common stuff of early design criticism in both Germany and England. The most remarkable difference was the relatively more relaxed character of the English writers.[3] This was to be expected from an organisation like the DIA which took a large part of its inspiration from the rustic musings of the Arts and Crafts fanatics. The elements of sound, austere commercial good sense which are fundamental to the Werkbund ideology were more muted in England, even if the problem of knocking into shape an industrial structure whose power was far in advance of its taste was more acute north of the English Channel.

There is in every early DIA publication the hidden spectre of massive industrial resources carelessly producing tacky or dated products in the vague hope that one note on an arpeggio of different stylistic options would resonate on a sensitive area of popular taste. Quality in consumer products was also felt to be poor. That these ideas and opinions are still familiar today is a testimony to how far advanced was the DIA's ideology in 1915. A glance in the direction of the Association's ancestor, the Arts and Crafts Movement, made its founder members recall that workmanship should be improved, as well as design, although the influence of the Werkbund made the acceptance of the 'fitness and economy' lobby of the machine enthusiasts inevitable.[4,5]

Like the Werkbund, the DIA maintained a dogged adhesion to 'functionalist' ideas, at least in so far as they were understood at the beginning of the century. The notion that 'black is an honest colour' is typically half baked but, viewed in the reflected glow of Edwardian vulgarity, seems to be an idea of prodigious novelty and astonishing potential. This too-ready assimilation of rules of thumb about the appearance of things – a belief that there was an absolute 'good design' in existence somewhere, if only it could be found – marked the early history of both the Deutscher Werkbund and the DIA. Increasing realism, brought about by the exposure of both organisations to the obdurate reluctance of manufacturers – other than those who were cranks or philanthropists – to listen to their arguments, modified the character of the DIA as it developed. During the 1920s and 1930s its activities were expanded to include the publication of the magazine *Design for Today*, the organisation of exhibitions, a slide library and continental excursions, and the establishment of a service to advise manufacturers on good design. The fate of DIA policy was to be absorbed into official government policy in Britain; depending on your point of view, it is possible to see this fate either as the crowning achievement of a just and successful crusade, or as the final ignominy of a lost good cause. Whatever the case, the existence of the DIA had not done enough by the 1930s to diminish clamourings for a

A PELICAN SPECIAL

ANTHONY BERTRAM

DESIGN

Front cover of Design *by Anthony Bertram, Penguin Books, 1938 (Photo Design Council)*

renewed attitude to design among responsible bodies in Britain.[6]

Yet if the health and vigour of a body of theory may be measured by a government's reluctance to act on its proposals then, for a time in the early 1930s, the DIA was still a healthy body. The turning point was Lord Gorell's involvement with a committee on art and industry, which he chaired. The Gorell Committee reported in the spring of 1932, intending to persuade the national government that annual design exhibitions, on the model established by the *Ausstellung* of the German Werkbund, would be a good idea to invigorate the lethargic body of British industry. Initially, the government refused even to vote a modest £10,000 required for the exhibition scheme. By the time Gorell became chairman of the committee which created the 'Design in Modern Life' exhibition at Dorland Hall in Lower Regent Street, which ran from June to July 1933, the government had relented and had given support to his efforts. The Dorland Hall exhibition was a microcosm of all the false hopes and misconceptions which have bothered British industrial design ever since. The exhibition concentrated on soft furnishings and glassware at the expense of real industrial design, although its more positive influence was to stimulate a barrage of publications to satisfy a public curiosity about design which had just been aroused. A series of articles called 'Design in Everyday Life' appeared in the BBC's new paper, *The Listener*, and was extended in titles published by Allen Lane's newly founded publishing house, Penguin Books; they were summarised by the official Council for Art and Industry Report of 1937.[7]

By the time Penguin Books started carrying advertisements for the DIA, the Association was beginning to look rather as though it lacked credibility, and the leadership in the design debate in Britain was passing into the hands of individuals like Nikolaus Pevsner,[8] John Gloag and Anthony Bertram,[9] who made it their own business to make the British public and British industry aware of their aesthetic responsibilities. In the United States the development of modern industrial design took an entirely different course. Never promulgated by para-official bodies, it was always in the hands of individual entrepreneurs who, locked as they were in a Darwinian struggle for existence with the dinosaurs of industry, forced the development of the most refined and exquisite science of designing for industry. It was in America that the very idea of freelance, commercial design offices – entirely independent of manufacturing industry – grew up. This phenomenon was the responsibility of a handful of men, pre-eminent among whom were Norman Bel Geddes, Raymond Loewy, Henry Dreyfuss, Walter Dorwin Teague and Harold Van Doren.[10,11,12]

It was these men, working as industrial designers of objects as diverse as duplicators and dumptrucks, who created the imagery of the twentieth century in a country whose avidity for image instead of substance has been so eloquently described by Daniel Boorstin and Vance Packard. Soon the manufacturing giants themselves were forced either to hire these men or to create their own design teams to do the job for them.

Typical of the almost mystical devotion which industrial designers in the USA gave to the fine detailing of their products was Raymond Loewy's development work on the Coca-Cola bottle, a famous shape if ever there was one, and which the company claims can be recognised by 90 per cent of the world's population. Loewy did not actually design it; that was done in 1915 by Alex Samuelson and T Clyde Edwards, but in the tradition of industrial design that he was to do so much to establish, he developed and refined the existing product (by making subtle alterations in the shape and adding white painted lettering) to create the bottle that is still in use today. His modifications to the Coke bottle made it – after the Cross, the Crescent and the Star of David – one of the world's first international symbols.

The handful of individuals who created the American industrial design profession established similar workshop practices and similarly persuasive arguments about the commercial necessity of good design. Each, however, worked from a rather different basis: Raymond Loewy had been an engineer, army officer and fashion illustrator; Walter Dorwin Teague had been a typographer and advertising draughtsman; while both Henry Dreyfuss and Norman Bel Geddes had worked in the theatre. Initially, each had to go cap-in-hand to industrial giants, offering the services they had invented and intended to provide: improved looks for a product, superior ergonomics (although not always) and increased sales. Each developed a personal technique: Raymond Loewy liked to lock himself away to aid his thinking; Norman Bel Geddes conducted consumer-use surveys; while Walter Dorwin Teague wandered in the valley of the shadow of philosophy and sought to extract the essential rightness of all objects and to express it in the designs he developed for them.

Of all the leading designers it was Norman Bel Geddes who described most precisely the processes involved in creation. He set up his studio in 1927 and from then on he employed anything from 15 to 70 staff designers in his private academy. In the case of one of his first commissions, to design the bodywork of the Graham-Paige car, Bel Geddes was given a tight brief by the client which restricted his imagination. In the year of this commission he was asked to envisage the Graham-Paige of five years hence and then to work backwards, providing a gradualised design for each model year. Bel Geddes carried out this rather deceitful task, but the design was not to Graham-Paige's satisfaction and it was never produced. More often, Bel Geddes was successful, perhaps because he always stressed that design was almost entirely a matter of thinking and that visualisation came last in the design process.

The procedure which Norman Bel Geddes adopted for his own design practice is an exemplary one. In considering product design he observed the following points:

1 Determine the precise performance requirements for a product
2 Study the methods and equipment employed in the client's factory
3 Keep the design programme within the budget
4 Consult experts on the use of materials
5 Study the competition
6 Conduct consumer-use surveys on existing products in the field

With this research done, a survey completed (perhaps conducted on a trans-continental train by Bel Geddes himself) and the object clearly in mind, the designer would go ahead and actually *draw* his vision of the product. It was, according to Norman Bel Geddes, this visualisation which was the last and quickest part and that which required the least work.

American product design during the twentieth century has been dominated by different temporary concerns, the most familiar and persuasive one being streamlining, an enthusiasm born out of

Coca-Cola bottle and fountain glass, 1915-55 (The Coca-Cola Company)

the popularity of commercial aviation in the years before the Second World War. According to Harold Van Doren, America first heard of streamlining in 1867, when a patent for a 'streamlined' train was filed by Samuel L Calthrop. By 1917, D'Arcy Wentworth Thompson was using the word in his book on form and by the 1930s, streamlining had become one of the obsessions of American manufacturing industry. The most celebrated American industrial product to embody the principles of streamlining was the Chrysler Airflow motor car.

The story of the Airflow is a moral tale representing the intrusion of industrial design into an industry which was not prepared to accept it; there ensued a débâcle rather than a triumph. The story begins in 1927 when the engineer Carl Breer and his engineering team at Chrysler started making experiments with streamlined forms for the dumpy Detroit products. It took seven years of cautious experiment, often against the will of the Chrysler main Board, but with the sympathy of the progressive Walter Chrysler, to produce the Airflow package. Walter Chrysler stands out, in a short-sighted field, as one of the visionaries of American industry. The Airflow, imitating on the land what Douglas Aircraft were doing in the air, was resisted as a concept by the cautious Chrysler executives. It was also resisted by the American public. It had been one of the first attempts which American industry had ever made to co-ordinate engineering and visual virtue in a single package which was honest and complete – and it failed. Norman Bel Geddes was called in to tidy up the front of Breer's design to see if his skills could improve the public's response, but even his visual ingenuity failed to make the dramatic Airflow catch on with the dull public. The Airflow's commercial failure burnt

Chrysler Airflow motor car, 1934 (Chrysler Corporation)

Chrysler's fingers and, although its appearance and its totality as a concept influenced the final form of Ferdinand Porsche's Volkswagen, it was the last attempt before the 1960s by the American motor car industry to satisfy both sophisticated and popular taste. When it failed the industry learnt not to do it again and, abandoning any lingering enthusiasm for wedding appearance with engineering, developed instead an attitude to motor car design which the Germans call 'Detroit Machiavellismus'. This aims only at immediate, temporary satisfaction. It is styling, and its development as a major component of the American design business is the other side of the coin from the earnest professionalism of the private design studios.

The role of styling in the American motor car industry is, at once, the most exquisite refinement and the most regrettable despoliation of all that has directed the attentions of enlightened industrial designers. Styling was the second way in which the American design industry followed in the years after the Second World War. Before the war, such product design as existed was, according to Niels Diffrient, just a judicious bending of metal in accordance with ephemeral aesthetic whims. By the later 1940s, however, firms like Henry Dreyfuss Associates, Diffrient's employer, had developed an approach to industrial design relying on ergonomics which was to carry them through the next two decades.[13] At the other extreme, the overwhelmingly successful firm of Raymond Loewy employed the master's unerringly accurate sense of style and of modern beauty to produce motor car designs whose influence can still be discerned 30 years later. It will be the job of future historians to determine which approach has in the long term proved the most beneficial for the future development of industrial design, but the stature of

Loewy's contribution to the appearance of the twentieth century cannot be disputed. Under his influence – to cite just one well known example – the conservative Studebaker Corporation of South Bend, Indiana, found themselves the astonished manufacturers of motor cars which represented the wild dreams of a man who, 20 years before, had been exhausting his imagination on the pages of fashion magazines.

Loewy's prowess in designing the Studebaker line combined the genius of the visualiser with the push and persuasion of a skilled businessman. Presenting some new designs for a future model to the Studebaker Board, Loewy first of all showed them a relatively sedate design for a saloon car, which gained their approval, and as soon as Loewy sensed that he had won the day he then unveiled another design which had been kept under wraps. This was the sensational coupé, the first American motor car with a European look, which carried the day for Loewy and his business. In styling it he had employed many clever tricks, including a sheet metal interpretation of the convertible's stowaway fabric hood. Loewy had found out that while most people liked the *look* of convertibles, relatively few soft-top customers went to the bother of stowing the hood away. His Studebaker coupé took advantage of this whim and the 'hard-top' was born, a car which combined the svelte appearance of a convertible with the practicality of a saloon. By 1953 Loewy's Starline coupé was so successful that it accounted for 40 per cent of Studebaker's sales.

In this consumer part of the American market, symbolism has always been an important and necessary part of any undertaking. Loewy used European, convertible symbolism for Studebaker, while Norman Bel Geddes, for instance, when working for the Autocar truck company, had tried to build the look of solidity – which a rival's heavyweight line had used to dominate the market – into the Autocar company's medium-weight range. But to Harley J Earl, chief wizard in the den of kitsch that was American car styling during the 1950s, symbolism meant employing just any motifs he felt appropriate at the time. Using a repertoire of imagery that filled the modest gap between Flash Gordon and the USAF, Earl created some of the most extravagant *and* influential product designs of the century. His career was an example of the power and authority which American manufacturing industry might give to a designer employed as a staff member: for a whole generation, Earl dictated almost every mode, idiom and genre of car styling.

Among the visual innovations that Earl developed were the wrap-around windscreen and the sculptural use of chrome. One of the ideological innovations for which he will always be remembered was the single-handed refinement and development of the credo of car manufacturing: planned obsolescence. This policy of making last year's product look démodé by prematurely introducing next year's ever more startling 'designs' maintained the American industry in a constantly rising helical progress of increased demand and increased production until the oil crisis of 1973 forced a reconsideration of the ethic.

Earl had had a free hand to create car body designs of the purest whimsy in a market which had always employed conservative engineering. Aimed largely at the home market, Earl's employer, the General Motors Corporation, had established by the 1940s a formula which was popular with the public. The early refinement of the electric starter had established the use of large engines in American cars which the consistent

'A break in the traffic' by Philip Castle (Thumb Gallery)

Buick Le Sabre 'dream' motor car, designed by Harley Earl (at the wheel), 1954 (General Motors)

Lockheed Lightning P-38 warplane, 1939 (Imperial War Museum)

availability of cheap petrol did nothing to undermine. A long wheelbase and a heavy engine ensured a good ride and this, in engineering terms, was what the public wanted. It was Earl's job to satisfy their fantasies as well.

He explained himself fully in an interview with the *Saturday Evening Post* in 1954, voicing the opinion that the General Motors Styling Department, which was effectively his sole creation, was devoted to pure form – a philosophy from which the cautious General Motors engineers diverged only when the styling department proposed an unstable three-wheeler with a cyclopean headlight.

The commercial necessities of American industry forced certain procedures on Earl. He was an enthusiastic follower of motor racing and its imagery was a constant source of reference to him. He was also interested in giving, say, an iron-engined, leaf-sprung barge-like Oldsmobile the appearance of high technology. His main visual idea in all his career was to make cars look longer and lower, in appearance if not in reality. His planned obsolescence policy forced him to visualise today what the public would want in three years' time, while what he had designed three years ago was just then rolling off the lines. Earl's achievements are interesting because in them we can see the evolution of visual tricks and the assumption of a vocabulary of imagery which have been immeasurably influential in purely visual terms, but which bear no relationship whatsoever to the lofty canons of High European Design that historians often like to emphasise. The necessity to stimulate the market and forever to produce something newer and yet newer still was a cruel tax on the imagination. It did not bother Earl at all; he maintained throughout his career an entirely unself-conscious attitude to his design procedure. Of planned obsolescence he said

'For one thing, we know that you car buyers today are willing to accept more rapid jumps in style than you were 20 years ago.' The pursuit of the new became a compulsion and Earl became a seer whose prophecies were self fulfilling.

New materials, new imagery, new conceits and new deceits characterised Earl's work. He would, he said, have used brass instead of chrome, but a judicious trawl through public opinion in a Bel Geddes-type consumer-use survey by a General Motors team disguised as newsmen established that the public preferred chrome. Of new imagery, Earl's most famous creation was the tail fin, a motif he borrowed from the Lockheed Lightning warplane. Earl justified this wanton addition, applied first to the expensive Cadillac model and spreading eventually all over the globe, by saying with direct and breathtaking honesty that the tail fins on the Cadillac gave 'visible prestige' to its owners. In these terms, Earl knew every trick. For the 1953 Cadillac he wanted holes in the forward part of the rear bumper to aid brake ventilation, but when the engineers so improved the brakes that the holes became unnecessary, a painted simulacrum of them was maintained for visual effect. Innovating again, when Earl's lengthening and lowering policy made the roof of the 1954 Chevrolet Nomad stationwagon visible for the first time, Earl immediately wanted to decorate it, to make it visually interesting. Not a square inch of exposed GM metal should be unexpressive. What did he do? He grooved it.

Earl's design policy which he developed for General Motors represents one extreme sort of obeisance to the power of the visual. Another was IBM's, one of the other giant American corporations which has assumed design into its organisation as a part of manufacturing no less significant than product-planning or even production. During the 1950s, one major innovation in industrial product design was the assumption by paternalistic companies like IBM in the USA and Olivetti in Europe of a deliberate and considered policy of 'good design'.

It was the example of Olivetti's impressive design programme which first influenced IBM to employ designers. Thomas J Watson Jr was president of IBM, the largest data, systems and office-products group in the world. It did not have a design policy. The story goes that, stuck in a New York traffic jam with the architect Eliot Noyes, Watson found himself outside the famous Olivetti showroom where the firm's products were displayed like sculpture on plinths in a spot-lit, marble showroom. He told Noyes that he regretted the lack of identity in IBM products and found himself to have hired Noyes as the corporation's design consultant. Noyes in turn hired the graphic designer Paul Rand and the architect-designer Charles Eames, as well as a number of 'name' architects, to help him with his IBM design programme. While the Olivetti products which Watson had admired were each very individual, expressing the designer's idea of style first, and the company's by the way, Watson wanted the whole IBM product line to have a uniform appearance. This made manufacturing sense too, because of IBM's policy, whereby one particular 'computer' might comprise a printer made at Endicott, New York, tape-drives made at Boulder, Colorado and a central processing unit made in England.

Although in a high-technology field like data and systems there is little impulse buying, Noyes and his team were very conscious that even hard-boiled engineers have taste. Indeed, in some instances, their clients were so pleased with their new status-loaded computers that, in mute imitation of the Olivetti showrooms which had inspired Watson, they placed them proudly on public view behind plate-glass windows.

IBM has built up an immense experience of client design preferences. Colour was one of the first considerations, and the IBM design team has found that 70 per cent of the customers who ordered the popular Model 40 computer specified the colour now known internationally as 'IBM blue'. Second, IBM wanted an expression of the quality inherent in their goods to be given a visual character. This was achieved under the vast shuffling and refurbishing of a whole identity executed by Eliot Noyes. Although inspired by Olivetti, the IBM experience was peculiarly American in its scale and thoroughness.

The progress of industrial design in Europe since the Second World War has been quite different. As was the case earlier in the century, design in Europe has largely been the responsibility of para-official bodies, such as the Council of Industrial Design (later the Design Council) in England, and the Hochschule für Gestaltung at Ulm in Germany.[14] These organisations tended to perpetuate the ideas about 'good design' which had been established in Europe before the Second World War and were able to do so without any competition from an articulate, private-sector design lobby such as existed in the USA.

The Ulm Hochschule für Gestaltung has been a focus of discussion about the course of European design for more than two decades. It self-consciously imitated the example of the Bauhaus (1919-33) by promoting self-expression, learning-by-doing, re-education of the senses and a study of the practical applications of art, but it soon moved beyond Bauhaus dogmatism to one that

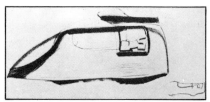

Study for Olivetti Studio 44 typewriter by Marcello Nizzoli, 1952 (Stile Industria)

IBM 305 RAMAC computer, 1957 (IBM Archives)

Packaging for IBM, designed by Paul Rand, c1960 (IBM Archives)

Poster for Olivetti by Xanti Schawinsky, 1935. (Stile Olivetti Kunstgewerbemuseum, Zürich, 1961)

Braun Phonosuper SK4 record player, designed by Dieter Rams and Hans Gugelot, c1956-57 (Braun AG)

was all its own. Partly by its association with the Braun AG manufacturing concern (for which Hochschule staff designed a whole factory in 1958), the school has become inextricably identified with a Germanic type of chilly good taste: admired at a distance everywhere, abundant in the suggestion of common sense, but undeviating in its lack of feeling or compromise.

The Ulm-Braun idiom is the most complete expression of the aesthetic convictions which were established before the Second World War and which, through the medium of the persuasive Modern Movement texts used in art and architecture schools everywhere, have become the new academicism. The novelty of the Braun idiom was that it was the first time that European manufacturing industry had so fully embraced a complete aesthetic. Braun took its new direction from Fritz Eichler and Hans Gugelot and later from Dieter Rams. The Braun designs they created dominated a *Der Spiegel* special feature on modern living as well as the 1955 Düsseldorf *Funkausstellung* where the new Braun designs symbolised the emerging *Wirtschaftswunder*.

The regeneration of the Braun image would not have been possible without the partnership of the Ulm school. With characteristic and telling hyperbole, Reyner Banham – chief rhapsodist of both the Modern Movement and the street-level popular culture which displaced it – has said that the Braun style 'in its abstraction and platonic idealism and aloofness and classicism relates to the most central and elevated concepts of established Western culture'. And he was right.

The Hochschule für Gestaltung was created by Inge Scholl, in memory of her brother and sister, who were shot by the Nazis. It was intended to be non-political and independent and Max Bill, who had written a book in 1951 similar in concept to this one, was called in to design the buildings; he became its first director and remained in charge until 1956, when a board which included Tomàs Maldonado, Otl Aichler and Hanno Kesting took over. Many of the teachers had been pupils of Walter Gropius and the school soon became known as the 'new Bauhaus'. Gropius had been consulted about the character of the original curriculum. Like the Bauhaus, students took courses in sociology, history and applied physiology, as well as developing a familiarity with materials and industrial processes. Given this knowledge, the young industrial designer was intended to be equipped to help manufacturers in design decisions. However, because – again like the Bauhaus – the student was considered an 'artist', many industrialists were not always so sure, and the Braun idiom which members of the Hochschule für Gestaltung created did not please everyone. When 'artists' like Hans Gugelot designed machines they appeared to be some sort of expression of a rarefied Absolute, bearing no actual relationship to the mundane duties of fan heater and food mixer. Gugelot exploited the easy association which the Western mind makes between a product being neat and clean and its being a 'good design' or a 'good machine'. Gugelot may not have been prepared to accept that his machines were *not necessarily functional*, but were abstract thought given sculptural form. That they were all beautiful was never doubted and that they were useful was never questioned but, as Reyner Banham again pointed out, what happens when you take the Absolute into the kitchen is that you get it dirty.

Braun, along with Citroën and Olivetti, represented the lofty apogee of European design in the 1950s, at least as far as it was understood by a sophisticated elite, writing for the architectural press.[15] Somewhere on the lower slopes of clique-acceptance was the popular Italian craze which dominated British taste in the later 1950s and which found expression in the vogue for products like motor scooters (which the racing driver Stirling Moss did much to popularise), Olivetti typewriters (often left unused, but doggedly sculptural), Espresso coffee machines and Gio Ponti furniture. Although many Espresso machines (even those designed by Gio Ponti) looked like Harley Earl's rejects, Italy became one of the cult homes of 'good design', a prejudice encouraged by the popularity of magazines like *Domus* and *Casabella* in which even non-Italianists could see glossily displayed the latest fantasies of Joe Colombo or the Castiglioni brothers.

In terms of indigenous design, Britain was as much at sea as it had been in the 1930s. The generation which, in Germany or Italy, was involved in developing industrial design was, in Britain, active in the art schools, riding on a high wave of enthusiasm produced by the rush of ex-service students. It was significant that the sort of art which the art school generation of the mid-1950s was producing either emulated or satirised the machine culture which had preceded it. This phenomenon, which has become known as Pop Art, included among its most eloquent celebrants the influential historian and critic Reyner Banham. It might serve as a temporary epitaph to British industrial design to note that the British activity in the visual arts which has attracted the most international attention has been Pop Art, which has used for its substance and its motifs the achievements in industrial design of Harley Earl, and others like him, whose aesthetic standards have been the very opposite of those supported by the 'official' vocalists in this country.[16]

The problems of relating art to industry which so exercised writers in the second, third and fourth decades of the century have still not been solved, least of all in Britain. The essential, intractable problem is that public taste is at odds with what a visually sophisticated elite prefers. This gulf is eventually being closed by the laws of scale which manufacturing concerns have to obey. The size of most efficient industrial undertakings tends to ensure a wider dissemination of successful designs than even the most dewy-eyed visionary could have imagined 40 years ago. This same scale of manufacturing, whose size makes 'mistakes' expensive, tends also to ensure the maintenance of a certain sort of quality.

The character and conditions of good design are still, however, a matter for debate. This book provides a visual and textual means for looking at problems which have bothered designers of all kinds over the decades during which an identifiably modern kind of industrial design has emerged. At least three – maybe more – basic patterns emerge, but all the designs illustrated here have one thing in common: by visual means alone they were meant to please. Until all manufacturers attempt to do this it cannot be said that the austere ideals of the Bauhaus or the Hochschule für Gestaltung, or the extravagant vulgarity of Harley Earl, or the shrewd business sense of Raymond Loewy or IBM need to be replaced by an alternative.[17] And although no concept of good design can any longer be said to exist, at least those companies who have employed industrial designers and remain both visually and economically in good shape, can testify that a faith in beauty is never in vain.[18]

The reference numbers indicate texts in the section beginning on page 25

1 Banham, 1955 (page 77)

2 Lux, 1910 (page 25)

3 Clutton Brock, 1916 (page 28)

4 Behrens, 1922 (page 32)

5 Ewald, 1927 (page 27)

6 Morton Shand, 1930 (page 41)

7 Council for Art and Industry, 1937 (page 48)

8 Pevsner, 1937 (page 52)

9 Bertram, 1938 (page 58)

10 Bel Geddes, 1934 (page 45)

11 Teague, 1940 (page 62)

12 Van Doren, 1940 (page 67)

13 Dreyfuss, 1955 (page 82)

14 Russell, 1949 (page 73)

15 Barthes, 1957 (page 88)

16 Hamilton, 1960 (page 90)

17 Loewy, 1945 (page 71)

18 Kaufmann, 1960 (page 95)

A Bicycle is Beautiful

Joseph-August Lux was born in 1871 and died in 1947. He joined the Deutscher Werkbund in 1908 and edited an avant-garde journal called Hohe Warte *which represented the views of the Werkbund on architecture and design. His other writings included* Die Moderne Wohnung und Ihre Ausstattung *(1904) and a book on Otto Wagner (1914). Towards the end of his life he wrote novels and musical biographies, but this extract is from his best known book,* Ingenieur-Aesthetik *(1910). In it Lux expresses all the naive wonder which a Utopian impressed by the possibilities of the new machines might be expected to possess. His views on the beauty of machinery are typical of the Werkbund ethic, as is his moralising imperative that all that is not expressive of its true nature is a sham. Although Werkbund apologists occasionally took a hard line against* art, *Lux is at pains to make clear that the imagination and conceptualisations involved in organising simple elements in a novel way is more intellectually challenging than the mere juggling with traditional motifs.*

Written at just the same time that Gropius' and Meyer's now famous Fagus factory at Alfeld was taking shape, Lux directs his argument towards architects. It is they, he suggests, who have most to learn from the example of machines; in them are all the elements of the new style.

The technical and stylistic achievements of modern times are revealed most clearly in contemporary machines. Even the older civilisations developed machines and the simple tool is essentially a machine. However, what distinguishes this modern machine age from earlier civilisations is the fact that the present takes its distinctive artistic or stylistic character exclusively from machines. Formerly this was different. The machines themselves used to be ornamented in the styles of the decorative arts. They were considered to be domestic utensils which should be embellished. A tool, a gun, a piece of baroque furniture, all bore the ornamentation of their age. Despite the use of machines, the products of earlier ages were handiwork. Now even handiwork is machine made. There is no product for which the machine does not take the bulk of the work . . . and whose form is not determined by the machine from the outset. As always, the transition which took place in the mid-nineteenth century from the style of the craftsman to that of the machine was characterised by the fact that the technical innovators continued to use the old, rigid form-language before summoning up the courage and determination to make use of their own specific modes of expression. The machine determines the styles of good taste which follow the principle of practicality. This forces the designer to consider from the start the metal hands which transfer his thoughts from the plan on to the material itself. Technology has gained precedence over art which it has forced into new channels. The new concepts of beauty, a new aesthetic, must be infused into technology, elevating the concepts of practicality and functionality to the highest principles. In Munich there is a museum of technology which presents valuable information on an area of which the artistic potential has not as yet been fully appreciated – the technical inventories and implements of the past. It includes background material to the machine age and demonstrates that the new aesthetic has always existed in embryonic form. Once freed from consideration of the historical factors determining style, it doesn't take us a moment to discover in these objects art forms, which have developed almost as a result of biological necessity. Most striking is the practical style of modern vehicles which shows the influence of the new machine aesthetic, as does that of the new large-scale constructions.

The question is whether an object is also beautiful because it is functional. Who would not consider beautiful a splendid vehicle, built lightly but solidly in mahogany or walnut, ingeniously constructed . . . comfortable and thoroughly practical? It satisfied our aesthetic sense to the point that we could wish neither to add anything to nor take anything from this object for perfection. True, the state coaches of previous centuries were well provided with decorative fittings which were not, in fact, an intrinsic part of them. Some features of Spanish court etiquette can be traced in ceremonial court appearances even to this day. But since we ourselves no longer wear the full-bottomed wig we have no reason to choose such decorative accessories as the standard by which we measure beauty. For we should not forget that the fundamental appearance of even these sumptuous carriages was determined by the style of construction which came right to the fore in the nineteenth century, the form of which is primarily adapted to the simplicity and practicality of our person, our clothing and our household effects as they have evolved through indigenous tradition. On the same basis a

bicycle is deservedly described as beautiful. A car, an express train, an excellently fitted-out railway carriage, a steamship and a speed boat are also beautiful. What more can be said about a rowing boat in this context? The same applies. Let us consider an eights boat: a light, narrow frame of cedar planks, about 15m long with wide outriggers for oars or sculls, sliding seats on wheels or runners so that the work of the leg muscles can support the arm muscles while rowing – a highly ingenious organic co-ordination of effort on the principle of economy which not only prudently saves exertion by increasing impetus and decreasing frictional resistance, but at the same time enables greater efforts to be made. This practical way of thinking has developed vehicles into structures which are almost endowed with human life. Our vehicles incorporate elements of our nervous system. It is a fallacy that functionality leads to impersonality. Quite the contrary. It . . . allows for extensive personal differentiation. Between the most primitive and clumsy canoe and the sensitive and manoeuvrable rowing boat there lies an enormous amount of mental effort, which imprints the human physiognomy of objects with regard to the level of our intellectual capacities. Our civilisation is reflected not in architecture, but in vehicles and modern transport engineering. If we are looking for the style of our age we will discover it here. A specialised art has developed, suddenly expressing our real nature. That carriage building represents a portrait gallery of separately developed individuals can hardly be ignored . . .

The motor car industry had developed numerous types of motor vehicles in consideration of the most diverse personal requirements. To give a rough outline; the motor car unites three important independent developments – the formal nature of

Fagus factory, Alfeld, designed by Walter Gropius and Alfred Meyer, 1911 (Photo Tim Benton)

German omnibus, c1912 (Bildarchiv Preussischer Kulturbesitz, Berlin)

coach building, the bicycle's principle of the sprocket-wheel gear and the dynamic force of the engine as a welcome substitute for animal muscle-power. The highly advanced tradition of coach building stands the motor car in good stead as regards aesthetics. On this basis it was possible for the motor car to develop a form which can already be considered faultless within a short space of time. Here we see again how formal aesthetic development arises out of practical necessities exactly as in the case of all other means of transport. The first motor cars still gave the unsatisfying impression of carriages from which the horses had been unharnessed. However, the motor vehicle soon got the shape which it had to develop in the interest of speed and the overcoming of resistance from atmospheric pressure. This was achieved by putting the engine in a narrow construction at the front end and by giving the whole structure the shape of a rapid locomotor like a bird, fish and ship. The airships too were forced to adopt similar shapes, which, although the product of necessity, could still be considered completely aesthetic. The pattern of the steerable airship had long been close to a formal solution: for its effective realisation all that was required was that the engine should be sufficiently light in the air so as not to be out of all proportion to the lift . . . .

The construction of the modern express train is determined by the same formal considerations. At the Nuremberg exhibition of 1906 an express train designed by Maffei was exhibited and not only technicians, but also aesthetes with a grounding in art, had to fall in love with it. It was built to reach speeds of up to 150 kph so that the driver's cabin, the smoke arch door and the panelling of the steam-dome and the exterior cylinders were designed as wind breaks. The side view of the front

section resembled the breast of a bird. It also brought to mind the streamlined construction of a ship or a fish's head. It looked as though it was designed to fly, even if only on the flat. Its intended function was so obvious that it needed nothing more to be beautiful. The way from the . . . design of the first locomotive to the perfected practical version of today marks a progression by which we have only gained. Thus new concepts of beauty have evolved in our age, concepts which can in essence be attributed to the idea of harmonious practicality and functionality. We cannot possibly call an object beautiful, no matter now lovely the ornamentation, if it totally fails or only partially succeeds in fulfilling the function for which it was intended. By comparison the expression of honesty and solidity is always pleasing and the more perfect and pure this expression, the more satisfying it is to our sense of beauty. Moreover, quite often an ethical principle is involved and this rejects the lie and masquerade and is, in the final analysis, an important factor in art and the appreciation of it.

The predominance which engineering and transport technology have gained before all else in the public interest can be largely explained by the fact that this field continues to be the object of much thought and the focus of a struggle for progress. The cartwrights or coach builders, bicycle designers, boat builders, mechanical engineers, car manufacturers and ship builders have all exercised their minds to a far greater extent than can be said in the case of architecture. It is considerably easier to play around arranging traditional stylistic motifs than it is to set about an intensive exploration of those requirements of life which are as yet undeveloped, as is the case with modern technology which has, indeed, enriched human life by a vast number of wonderful inventions. Technology has

Zeppelin airship (Die Neue Sammlung, Munich)

Arthur Clutton Brock

not only extended the field of our knowledge, but also the range of our capabilities, in short, the sphere of human influence, and has lent us powers which only 50 years ago were incredible dreams.

So it is here, in the field of technology, that the seeds of a new architecture are to be found. For, in the final analysis, technology deals with the establishment of contacts with the natural world beyond us, the extension of the sphere of influence of our organs and nerves. We want our voices and arms to span the ocean, we want to unite countries, to shrink distances of time and space by means of cables, express liners, motor vehicles, multifarious means of transport, by the construction of railways, bridges and tunnels, by using discoveries of all kinds as long as their form is the product of necessity, is determined by practical factors and is not encumbered with any preconceived notion of style which belongs to the past. This is life. A new concept of space and design comes into being, a new concept of architecture, a new concept of beauty.

From 'Maschinenaesthetik und Verkehrsmittel' [The Beauty of Machinery and Vehicles], in Ingenieur-Aesthetik Verlag von Gustav Lammers, 1910, pp 50-56, translated from the German by Sarah Golding.

Art, Workmanship and Style

Design criticism in Britain has often concentrated on two themes. The first is lamentation of the crudity of public taste, the second is condemnation of the lethargy of industry. This essay by Arthur Clutton Brock (1868-1924) combines both. Brock was trained as a barrister, but soon turned to literature and the arts, becoming an editor, belle-lettrist and, ultimately, art critic of The Times. *He was also an active member of the DIA, for whom he wrote promotional material. This is an extract from his paper,* A Modern Creed of Work, *which was published as the fourth pamphlet of the DIA in 1916, the year after its foundation.*

Brock eloquently represents the credo of the DIA, whose beliefs were essentially similar to those of the Deutscher Werkbund, but tempered by traditional English restraint and the influence of the same sort of low-church moralising which affected a part of the Garden City movement. Another difference between the Werkbund and the DIA was that the German organisation had very real links with industry, while the English one was still essentially craft oriented (having developed in direct line of descent from the Arts and Crafts Movement through the Art Workers' Guild).

This fragment of Brock's writing has been chosen as typical of a particular sort of English design criticism which was to develop during the years before the Second World War. Much of what Gloag, Bertram and others were writing in the 1930s owes its inspiration to these early pieces published by the DIA.

The word art does not occur in the title of the Design and Industries Association, because art is commonly supposed to be a mystery understood only by artists; and the Association wishes to make no mystery about its aims. The word design has a more precise meaning than the word art. When we say that anything is well designed, we mean that it is well fitted for its purpose. To design a chair well is to fit it to its purpose as a thing to sit upon; and where there is not good design there cannot be art. But in all objects of use we have made a vicious distinction between works of art and other things. We think, commonly, that an object is a work of art only if it is ornamented; for we suppose that beauty consists in ornament. But ornament only makes an object more ugly if it is not well designed and also well made. What is needed before there can be beauty in any object of use is good design and good workmanship; and the aim of the Association is to improve the design and the workmanship of all objects of use.

How little we recognise that beauty depends upon design can be shown by one notorious example. The Tower Bridge, while it was merely a great work of engineering, had the beauty of fitness. But this beauty was thought to be merely naked ugliness; and an architect was employed to clothe it with art. Therefore he concealed the piers of the bridge with two Gothic towers, which destroy all its beauty of design and make its strength look like weakness. So we are always destroying the beauty of objects of use with what we call art; and we do this so constantly that we have, for the most part, lost the faculty of good design or the power of recognising it. Beauty to most people consists, not in design, but in what they call 'style'; and style changes as quickly as fashion in dress. Thus, people get a notion that high finish is inartistic, as it is when it is finish for the sake of finish; they suppose that there is some mysterious virtue in the roughness of peasant art; and they will buy objects in which this roughness is imitated for commercial purposes, objects that are merely badly made. But where there is no question of art no one will put up with bad workmanship because of its style. Since, for instance, a motor car is not art to us, we exercise our common sense upon it; we do not accept bad finish in it because we want it to look like a peasant motor car. We wish our cars to be both well made and well designed; and so they are more beautiful than most of our art. But if we wish our art also to be well designed and well made, even that will become beautiful.

Good design and good workmanship produce beauty in all objects of use. That is the common sense of the matter. But human beings never attain to common sense unless they aim at something beyond it. There must be a kind of religion of workmanship, if workmanship is to be good; and a religion of design, if there is to be good design. It never is good unless both designer and workman do their best for the sake of doing it. What we need most in England now, is this religion; and we need a condition of things, a relation of all the parties concerned, in which it will be possible to do good work for the sake of doing it. When we have that, we shall have art soon enough. And it is not an impossible or unnatural relation, for it has often existed in the past.

The delight in doing a job well for its own sake is just as natural to man as greed or laziness or fraudulence. There is a natural force in him making for good work, as there is a natural force making for bad. Unfortunately the force making for bad work is helped, at present, in England, by circumstances which can be overcome, and by a body of mistaken opinion which can be refuted. But the circumstances can be overcome only if the opinions are changed. Thus, both manufacturers and shop-keepers often believe that they are utterly at the mercy of the public taste, and that the public taste is quite irrational; the public does not want good design or workmanship; the only way to success is to tempt it with continually changing fashions. Unfortunately such beliefs become true, if acted upon, in trade as in politics. The public can easily be demoralised in both cases. It has not a fixed and certain taste of its own. It does not know what it wants, but is subject to suggestion; and if it is beset with articles ill made and ill designed but folloing some new and violent fashion, it will come to believe that these are the articles which it wants. Tradesmen, like politicians, can be demagogues, and can make their fortunes by demagogy. But there is promise as well as danger in the fact that the public taste is plastic. The mistake in England has been the belief that it is plastic only in one direction, or, rather, the belief that it is not plastic at all, but always in favour of plausible rubbish. Producers think they are giving the public what it wants, when really they are forcing upon it what they think it wants. The fact is that they can force upon it what they choose to give it. This is not true of the individual producer. He probably is not strong enough to withstand any general tendency of the mass of producers; but still it is the tendency of the producers that controls him, not the tendency of the public. So producers in the mass can control their own tendency, since they can persuade the public that it likes what they choose to give it. Therefore the question is whether they shall blindly, and without any forethought or organ-isation, submit to a general tendency imposed upon them by the worst among themselves, or whether they shall exercise their will in combination to per-suade the public that it likes what is good.

The future of all English industry depends upon their decision. In the long run a national industry is made powerful by its excellence, and it will over-come a national industry less excellent, in spite of all tariffs. But the excellence of an industry depends, first of all, on the religion of the workers, on their religion or workmanship. It will excel if they do not produce merely to sell, if their first aim is to make the best article possible. For that alone can give enterprise, invention, the sense of adven-ture, which are necessary to success in commerce. When the aim is merely to sell, the workman, the designer, even the middleman, is bored with his work. He thinks of nothing but the struggle for life; and life itself has no significance for him. No man is living well unless he feels the significance of life in his work and not merely in his pleasures; and the community which is not living well will be over-come by the community which is, overcome in material things no less than in spiritual.

Where an industry is dominated by the aims of its worst members, where it sinks to their level and makes no effort to raise them, it is a declining industry, and at the mercy of a foreign industry which dominates the worst with the best. There-fore it is necessary, even for purely material reasons, that an industry should organise itself against its worst members. The individual pro-ducer of shoddy may grow rich, but he is killing, for posterity, the goose which has laid the golden eggs for him; he is, in fact, a public enemy, a traitor to his own trade; and if his trade cannot now deal with him as it would have done in the Middle Ages, it can at least organise itself against him; it can teach the public to shun him by example as well as by precept.

It is the aim of the Design and Industries Association to produce such an organisation. Experience has shown that this cannot be done by artists alone, by manufacturers alone, or by shopkeepers alone. Experience in Germany has shown that it can be done by a combination of all three. But the aim of this combination must not be merely to capture German trade. Motive in such matters is a point of first importance, and a merely competitive motive will not produce excellence of any kind. Nothing will do that but the desire for excellence. The regeneration of English industry and art must be based, not on a fear or hatred of Germany, a motive transient and in itself evil, but upon a desire for good work, a motive lasting and in itself good. It must be understood that we have made a fatal mistake in our whole conception of industry, a mistake which is robbing both rich and poor of joy and enterprise and vigour. We have believed that competition is the soul of trade. It is not the soul of anything; and trade cannot live without a soul... When an enemy has a noble lesson to teach, it can only be learned from him nobly. The aim of the Deutscher Werkbund was not to overcome English industry. It has won its victories over us because its aim was the excellence of German industry. So the aim of the Design and Industries Association is the excellence of English industry; that and nothing else. This aim cannot succeed without co-operation between artist, manufacturer, and shopkeeper. Each of these classes is powerless without the help of the others. Each individual member of them is powerless against any downward tendency of his own class. And without co-operation, as experience has proved, the tendency will inevitably be downwards. The producer of shoddy and the seller of shoddy will drag others down to their own level, and the public taste with them.

From A Modern Creed of Work, *published as the fourth pamphlet of the Design and Industries Association, 1916. Reprinted with the permission of the Design and Industries Association.*

Peter Behrens 1922

'Style?'

This essay by Behrens is essentially a philosophical rather than a practical one and in this respect contrasts vividly with the discussion on locomotive design by Curt Ewald (see page 36). Written three years after the foundation of the first Bauhaus for the Werkbund journal Die Form, *the great architect and AEG designer used this essay to advertise his enthusiasm for pure craft as well as to come to terms with his belief in the Zeitgeist, or the spirit of the age. There is an almost Expressionist tone to Behrens' article. His tone was partially imposed by the drama of post-war circumstances in Germany and its mystical style is reminiscent of the Utopian (and revolutionary) dreams of architects who gathered together into the Novembergruppe or the Arbeitsrat für Kunst.*

Like the more mystically inclined Expressionists, Behrens – despite his well known connections with industry – does not have unmixed praise for the achievements of engineers. He paints a picture of them getting more and more carried away by their technical achievements, sometimes at the expense of the appearance of products they have created.

Behrens, like English designers and Joseph-August Lux in Germany, admired contemporary transport, especially road vehicles, but deplored those areas of contemporary industrial design where the push of technological development was less strong. In this highly sophisticated piece it was Behrens' intention to explore the higher meaning of technology. He ends with a refutation of mechanistic architecture and an attack on the ugliness of some modern art. Although Behrens obviously sympathised with the more Utopian ideals of the Bauhaus, these comments of his might be interpreted as critical of actual Bauhaus production during the phase when that institution began to pass out of its early Expressionistic period.

The creative artist – creative in the sense that he genuinely produces something new – is not one to concern himself with the style of his age. He works on what he wants, and leaves other things unfinished. Indeed, only too often he destroys what he has just begun. Yet even he cannot entirely escape the influence of his education, and is sometimes concerned whether or not his work will appeal to popular taste. Ultimately, though, intellectual reflection is unimportant to him. What matters is that his work should look right: he is indifferent to fashion and the cultural consensus.

This is the productive stance of the responsive artist. It is the person on the receiving end who enquires into causes and influences. Such queries are posed by the man educated in the arts whose background has been based on a sense of history and who realises that every age has its style, including our own. That we seem to lack style, and have done so for many years, is commonly accepted. However, is it not possible that a later era will be able to discern a common denominator in the great range of diverse artistic forms in our period, including those which we at present judge to be lacking in artistic merit? Already the museums have begun to collect furniture and utensils from the period of Napoleon III. A style is not recognisable in its own time: it can only be perceived at a distance.

It is therefore pointless to insist on defining the style of one's own age and inventing aesthetic pros and cons with respect to what is new and unfamiliar in the arts. A purely aesthetic approach will always remain superfluous, since the artistic will of a period will always go its own way, with the aesthetic approach trailing along behind.

Now, what most interests us and challenges our

intellect is our general human and personal relationship to the manifestations of life at large and to the structure of time. It is unthinkable that we should set ourselves apart from the events of our times and enjoy a romantic dissociation from all that goes on around us. And so we will participate in the happenings of our era and will fight the intellectual battle bravely or with bowed head, according to our temperament. Such a commitment is not an aesthetic but an ethical one.

No one is likely to underestimate the influence which the Great War and its consequences will exert on future generations. Our interim impression is that we have witnessed the collapse of a high economic civilisation, a collapse which not only affects ourselves, but all the countries of the Old World. The sophistication of this civilisation found its most typical expression in the unprecedented progress of its technology. It seemed then as though all endeavours were subsumed within the general perspective of mathematically-based thinking. And this view was reflected in the world about us. We admired the massive curves of those great iron halls and the bold sweep of those bridges: we succumbed to the aggressive impact of those machines whose construction seemed so audacious and so cogent, though we kept telling ourselves that such an impact on our senses was fortuitous and pseudo-aesthetic, given that no artistic conception underlay their construction.

And indeed technology has thus far contrived to reach a peak only in terms of material existence, for the unity of the material and the spiritual – that is, the values of culture – has never achieved formal expression. Indeed, nobody has even given a moment's thought to it. The engineer in his ascent from success to success has turned increasingly away from anything which has no immediate bearing on his own field.

Thus our times present on all sides the symptoms of disunity. Wherever we look, we see a welter of contradictory trends. We acknowledge the considerable accomplishments of the construction engineer, but we are equally away of the decadence and mediocrity of our modern city architecture. The bold forms of our transport vehicles are a constant source of pleasurable surprise, yet at the same time the quality of the general utensils and industrial products on display in our shops indicates a nadir of taste which can scarcely be imagined any lower.

This may prompt present-day intellectuals in particular to see technology as the medium and expression of a calculating, analytical and imperialistic era. And they may be inclined to turn back and seek out far-off times and places by way of an art of introversion and soulful simplicity. For them the machine has utterly destroyed the heart of the creative urge and the created work. They have a presentiment that our technological and materialistic civilisation is reaching its peak, and that a return to spiritual and cultural values awaits us in the near future.

Here we see the opposite tendency. Both cases are the same: both the engineer and the man of sensibility are opting out of a commitment to the totality of lived experience.

No one would seriously wish to give up the achievements of modern technology. More than at any other time in history, we are dependent on its greatest and cheapest facilities. Indeed, precisely because we are economically weak at the present time, we have no choice but to make our lives more simple, more practical, more organised and wide ranging. Only through industry have we any hope of fulfilling our aims. It alone can save us from our economic misery. But those people who object to seeing a split between mechanisation and the life of the spirit are equally in the right. It is thus a question of historic importance whether it is possible for technology to free itself from its role as an end unto itself and to become a medium for the expression of our cultural life.

How might this be achieved? Certainly not as a result of aesthetic principles exerting an influence on industrial affairs, nor through the lumping together of two essentially alien fields. The recognition of the higher meaning of technology is above all an attitude of mind. Assuming that those commentators are wrong who suggest that our era will see the end of technological progress and the beginning of a period of spiritual inward-turning, it remains to be seen whether technology will acknowledge that power brings responsibilities and whether it can come to terms with that fact. This is what all authority, be it secular or ecclesiastical, has had to do throughout the ages. In the long run nothing can endure if it lacks coherence. This means that it is superfluous to ask whether it is the engineer or the architect who will have the last word in the production of metal constructions and items of industrial origin. Here, as in all domains where organisational skills are important, everything depends on the working community. A very good proof of this can be found in the many excellent buildings that have been built in this way, which gives rise to the hope that our yearning for a synthesis of artistic talent and technological craftsmanship might well be satisfied.

Thus we do not believe in a defeat for technology, but rather in its transformation. Admittedly we might still fear that art might forfeit its freedom if it became dependent on the interests of workshops and sponsors. Our age may be rich in talent

and fresh artistic impulses may be forthcoming, but the works that have come about have not made an impact outside specialist circles and have failed thus far to contribute to a universal movement.

Contemporary art appears to us intent on demolition, on a conscious destruction of form. What used to be called Harmony or Proportion is no longer respected, whether it be in terms of construction, form or colour. It seems as if beauty has been replaced by ugliness. Who can tell us what beauty is? No term is more relative than this one: it has changed and keeps on changing from generation to generation. All we can hope to do is to try to understand the wider context within which all these instances are situated. We will then come to see that contemporary art is tackling enormous problems, and that it is determined to take part in global events. Even before the war, when nobody had any thought of revolution and collapse, the plastic arts had already engineered a wholesale revision of taste which now appears as a prophetic anticipation of the era that was to come. And now the point becomes clear: the content of this new art is nothing less than the search for a different and novel harmony, for a new structural order. Its nature is symbolic. It deals in abbreviations, in suggestive interpretations of an object viewed in general terms rather than from a particular angle or in the light of specific circumstances. This new art offers signs in explanation of its forms, thus adding a conceptual dimension. It is therefore not a picturesque sort of art, but a formalistic and conceptual one. All the same, its forms are not conventional, but are shaped by the consciousness of a greater coherence among all the things existing within the universe. However, this idea has not yet asserted itself in an external form capable of realising the concept of a fusion of the arts. The ideal of a uniform spirit which might inspire and unite all forms of art is still absent from our age.

And this too is not a matter of aesthetic but of moral awareness . . .

The idea of the fusion of all art forms must proceed from architecture. But the concept must not be understood as a mere conglomeration of the various arts, as might happen at an exhibition, nor as a decorative device which forces the arts into an artificial collaboration. It must be a working with each other and from each other, a kind of reciprocal dependence and support. These words might not sound novel, but they do represent something basically original in that they imply in the act of construction a manual involvement and total surrender to the materials conducive to the expression of organic growth. If this takes place, then all ornamentation will become redundant and indeed disruptive. Organic creation out of distinctive materials and the desire for decoration are fundamental opposites: they represent two contradictory attitudes of mind. One might almost be tempted to condemn as immoral all ornamentation which lacks symbolic significance, whether in relation to structural or other higher principles. And so we shall gladly dispense with the deft touch of talent and skill. Where it is necessary and appropriate, we shall welcome heaviness and severity in painting and sculpture, the free arts. For these too, as we have seen, are nothing less than modes of building, crafts in the best sense. In this way we shall easily succeed in bringing all the arts in their formal relations into close intercourse. They will support and regulate one another, with the result that we shall no longer understand how people could have made such a grotesque distinction as that between 'arts' and 'crafts'.

Only through a collaboration in the totality of our experience of the world can we create the concept of another, higher reality, just as life itself can only be fully understood through the sum total of our senses.

From 'Stil?' [*Style?*], *in* Die Form *(Berlin), Jahrgang 1, 1922, pp 181-184, translated from the German by Roger and Agnès Cardinal.*

Curt Ewald 1927

Thoughts on the Shape of Trains

Ewald's highly prescriptive piece is a pure expression of functionalist theory applied to industrial design. In it he not only gives voice to the credo that only technical necessities justify aesthetic adjustments, but also takes the opportunity to mention the benefits of standardisation, a principle at the centre of Werkbund thinking.

This essay, which appeared in Die Form, *was written in the days before aeroplanes had gained wide acceptance as good design born of technological pressure. At the beginning of the 1930s, an express train could still be considered the* ne plus ultra *of glamorous travel and an eloquent symbol of the century. By the end of the decade, with the appearance of the Douglas DC-3, Boeing 247D and the Lockheed Lightning, all that had changed and trains were merely curiosities which only rarely attracted the attention of designers or writers.*

However, Ewald's detailed arguments constitute a remarkable model of a certain attitude to design which dominated a portion of German thought at the time. He believes that adherence to the mechanical necessities will inevitably produce a beautiful solution to the problem of design, although, perhaps a mite nostalgically, he believes that this is only true of steam engines, because in electric trains the necessary mechanicals are best concealed. The authority with which he discusses the technical details of locomotive design and the apostolic arguments he forges from his observations make Ewald's essay a period piece of considerable interest.

If the form of a machine is stylistically pure, it is because it gives expression to its constructor's conscious intentions and actualises the plan he has conceived through the most perfect, simple and economically appropriate means. Its beauty is a reflection of its technical and economic perfection, and its formal aesthetic effect can be traced back to the thorough implementation of a functional purpose. The more clearly this notion is pursued, the more meaningfully will the design of a machine reflect the specific character of its function. Thus a machine may be deemed to be aesthetically satisfying when its function can be inferred from its appearance and is echoed in the shape and fit of each of its components. In this perspective, the interdependence of technology and aesthetics is reducible to a single formula such that even the layman is able to assess the design of a machine.

There is no doubt that in the whole field of mechanical design, the steam locomotive is the machine which enjoys the conditions most conducive to a satisfying aesthetic effect. Its basic overall construction is easily grasped by the observer: the boiler produces steam under high pressure, the driving power; this in turn propels the pistons in the steam cylinders, which transmit the power through the piston rods, drive shafts and coupling rods to the driving wheels and coupling wheels, thereby setting the locomotive in motion. The connecting link between the boiler and the driving mechanism is the frame, whose function is on the one hand to take up the weight of the boiler, and on the other to transmit the tractive power from the driving wheels to the draw bar. The locomotive is a self-contained unit, comprising the creation of power (the steam boiler), the transmission of power (the engine) and the release of power (the working parts). One's eye can literally trace the flow of power from combustion through the generation of steam, the driving mechanism, the driving wheels and the frame to the draw bar which pulls the carriages attached to it. Furthermore it can follow the vertical flow of power which derives from the transference of the heavy weight of the boiler onto the boiler supports, the frame, the axles and the wheels and finally the rails. Thus the primary features of the machine as a whole and those of its component parts are clearly visible: hence their overall organisation is an embodiment of the highest aesthetic values. Herein lies the considerable advantage of the steam locomotive as against other types of machine whose function cannot be immediately deduced from their external form. For instance, the stationary steam engine is separated from the source of its power and in large measure also from its working parts; and as for the electric locomotive or the machine tool, the eye is equally unable to spot the power source. The motor vehicle likewise keeps its mechanism hidden beneath a protective bonnet.

The production of massive power within the smallest possible space, the striving for high speeds and a kind of natural serenity associated with the regularity of the railroad track, these may be counted the major features of the steam locomotive, and they can be immediately deduced from its overall layout. These features find expression in the design in the emphasising of the boiler as the generator of power, in the rigorous exploitation of raw materials and of space, in the accentuation of the horizontal line running the length of the locomotive, and in a balanced organisation of contours and spaces. Adherence to these guiding principles leads inevitably to the following considerations.

The boiler must be emphasised by being placed high up within the total design. The long boiler is generally cylindrical: in certain cases a boiler may taper slightly towards the front, thus increasing the sense of forward impetus. The upright boiler with its ashpan is visually satisfying in almost every form found in modern locomotives. Where it immediately joins the long cylindrical boiler, it contributes to a general impression of serenity and balance. In the improved Belpaire design or when it is slightly raised up, the upright boiler is given an emphasis entirely befitting its position as the seat of the power source; a similar emphasis is achieved by the wide rear boiler placed above the frame. It has an advantage over the retracted firebox, which is positioned between the frame plates, in that its outline is clearly visible and its technical function easily recognisable from its shape. The diameter and length of the boiler are dictated by the use to which the locomotive is to be put. With goods engines and mountain locomotives, a short squat boiler is recommended, while with high-speed locomotives, the horizontal line is usually given more emphasis because of the greater length of their boilers. To be sure, boilers which are too long produce an adverse aesthetic effect, in that they indicate an uneconomical exploitation of the combustion gases. Either the smokebox has to be moved up in front of the cylinder plane, or else there is the danger that the cylinder, if it is brought too far forward, will weaken the impression of power given by the boiler. Locomotives with steam pivot mountings constitute in this connection a technically justifiable and hence aesthetically satisfying exception. The smokebox door, being the component which rounds off the front end of the boiler, is an essential element in the total picture, and the only adequate solution to its design is to opt for a curved metal plate, in accordance with advances of contemporary technology.

Superstructures above the boiler produce a more favourable impression the lower down they are situated – a realisation which fortunately corresponds to the demand for a high positioning of the boiler. Box-shaped superstructures clash with the round shape of the boiler; conversely, saddle-shaped or even round structures appear to grow organically alongside it. The steam dome should have a modest cap with only a slightly curved top. If it and the sandbox are hidden beneath a single cover, care must be exercised lest this cover weighs too heavily on the boiler as a result of its extreme elevation, or appears excessively short in relation to the length of the boiler. No set rules need be laid down for the design of the smoke stack, which can be fashioned in almost any style. Visible spark arrestors require particular attention. Those which give a pear or skittle shape to the smoke stack, or which fit around the neck of the smoke stack like a collar, can have no pretentions to being well designed.

The steam cylinders must lie horizontally, and must be strongly emphasised as befits their importance. Their functional significance dictates that the cylinder and the valve box should be disposed visibly and the the valve box should not be hidden inside the frame or under the metal running board. The cast-iron cylinders or the powerful frame braces between the cylinders offer a natural base for the forward boiler supports. Particular attention should be given to the way these boiler supports are worked out, for they dictate the way the eye will trace the flow of power from the boiler to the frame. Only the Americans have managed to design them in an aesthetically pleasing way, and

they do so by offering them to the eye without any covering at all, while at the same time furnishing the short thick steam pipes on the outside of the smokebox with a cylindrical casing to protect them against loss of heat. Our practice, on the other hand, is to hide the steam pipes behind a metal screen which conceals the boiler supports and thus gives the impression of being a boiler support itself. An illusion of this kind cannot be fully satisfying. All the same, it does seem to work, provided the screen is mounted directly over the cylinder in a suitable size and sticks out a fair distance from the smokebox.

The driving mechanism and the controls realise the transmission of power between the cylinders and the driving wheels. It is therefore only natural that they should be set up on the outside of the wheels. They look even more impressive if the individual rods are set in a horizontal or just slightly tilted position. And yet even locomotives whose driving mechanism is invisible (because it is hidden inside the frame) can be considered free from reproach, for their aesthetic appeal lies in the utmost balance and unanimity of their overall effect (eg the older English locomotives).

Wheels should be fitted with an appropriate number of spokes of elliptical cross-section. Burnished or white-painted tyres consolidate the impression of smooth motion and may occasionally offset the disadvantage of spokeless (whole-cast) wheels. The balance weights should fit snugly into the circumference of the wheels, and, in accordance with the theoretical determination of the most appropriate form, should be sickle shaped, but never wedge or segment shaped.

The frame should be positioned between the wheels, and only in exceptional cases can an external frame be technically and therefore aesthetically

Polish State Railways locomotive, manufactured by Hanomag

German locomotive, 1846, manufactured by Hanomag

Spanish Northern Line locomotive, manufactured by Hanomag

(Photos Massey-Ferguson-Hanomag Inc, Hannover-Linden)

justified. Technical considerations will likewise dictate the choice of either the closed plate frame, with its balanced effect, or the open bar frame, which accentuates the transmission of power in a dramatic way.

There can only be one principle determining the design of the cab; spacious form conducive to a well balanced effect. Graduated borders beneath the side panel, box-shaped ventilators upon the roof, barred windows: these create a disruptive impression. High arched roofs and ventilators spanning the whole length of the roof are recommended. Side windows should be shaped as a rectangle with rounded corners; angled corners give a jagged impression, while bow-shaped upper rims set next to the windows produce a disruptive effect. Only the one demand of serenity of effect is to be observed in relation to the running-board, which forms an important transition between the boiler and the driving mechanism: it should lead horizontally and without any variation in level from the cab until it reaches the forward buffer platform. The water boxes and coal boxes on locomotives which carry their own fuel (ie tender engines) must be situated in such a way as to increase the overall massive look of the machine. They must on no account diminish the emphasis on the boiler. The danger here lies with water containers which straddle the back of the boiler or with side boxes which are set excessively high up and in front of the smokebox.

A separately attached tender should complement the overall effect of the locomotive through the unity of its mass, and equally should not seem too small in relation to the locomotive. Here again, great stress must be laid on the value of well balanced design. Disruptive forms are only acceptable if . . . they are technically justified.

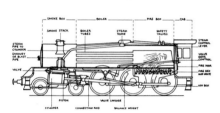

Schematic drawing of steam locomotive showing mechanical parts (Peter Hodges/Design Council)

Accessories and equipment have, as a result of thoroughgoing standardisation, developed into products with the highest technical fitness for purpose, and have thereby achieved a conspicuous formal beauty. Even so, care must be taken to ensure that they are usefully integrated into the design at large. Headlamps with projectors should be cylindrical and not box shaped. Number plates are flat and are only suitable when set against larger flat surfaces, and not against a cylindrical smoke stack which they can only touch in one section. Pipes, ducts and handgrips must wherever possible be set absolutely horizontally: they constitute an extremely valuable and simple means to accentuate the horizontal line and thus enhance the aesthetically satisfying impression given by the locomotive.

Where the individual components are designed in accordance with the above principles and without aesthetic faults, then their joint effect – the overall design of the locomotive – must necessarily evoke a sense of formal beauty. For the relation of the components to one another is laid down in accordance with a single and unmistakable technical principle, namely: the disposition of weight over the axles in the light of the operating conditions involved. Only in cases where the underlying technical premises are obviously behind the times can the propriety of a design become questionable. (For instance, bridges with a minimum load capacity make it necessary to set individual axles at abnormally great and thus aesthetically displeasing distances from one another.) In assessing design, one must therefore take account of the use to which the locomotive is to be put. Express locomotives require high driving wheels, and a forward pivot mounting for safe travel along the curves of the track, and no masses are allowed to protrude beyond the front or rear axles. All these requirements can be dismissed in the case of locomotives with lower speeds, without technical expediency and the concomitant aesthetic effect being thereby overlooked.

As can be appreciated, success in designing a locomotive built with a particular purpose in view comes as a matter of course. The external form is a consequence of an intellectual commitment to a task and in all logic will prompt a positive inference as to its technical prowess. Any ugliness in the design is an index of a fault in its technical functionalism (to be sure, a locomotive which has formal beauty need not be technically perfect); it is therefore completely aberrant and pointless to try to enforce an aesthetically satisfying formula by artificial means. A locomotive may possess formal beauty without its smoke stack being set above the centre of the cylinder; smokebox and cab doors which are pointed and streamlined are undesirable, for the cost of production stands in inverse proportion to the savings on fuel made by the reduction in air resistance. The high positioning of the boiler, with low fittings set at the same height, the emphasis on the horizontal line and the well balanced layout of the cab, the running-board and the fuel containers: these are factors of such paramount aesthetic appeal that they completely overshadow any minor inadequacies. They constitute the principles of design underlying the construction of massive smart locomotives whose attractive general impression cannot be denied; even if the superheater, say, is set somewhat inorganically in the smokebox at an angle conflicting with the horizontal emphasis along the length of the engine, or again if improvised wind deflectors are stuck alongside the smokebox. The undeniable aesthetic appeal of the locomotive will not be diminished by an awareness of the lack of expediency and consequent ugliness of the ancient side-buffers on the locomotives of our national railway. The smaller and the very small components exert a decreasing influence on the total effect: their design is motivated in the light of considerations of economy and simplicity of manufacture which are only indirectly related to function and purpose in the locomotive.

The above considerations are by no means novel: they have already served as guidelines in the design of locomotives from an earlier period. The difference between then and now lies simply in the fact that the notion of the 'high' position of the boiler was at one time interpreted differently. Furthermore, the disposition of the axles has altered in accordance with technical developments; to this may be added the previously inconceivable massiveness and sheer strength of many of our modern locomotives. The steam locomotive in its present-day form has reached a level of perfection upon which it is scarcely possible to improve. Future developments must lie – to the extent that one does not turn to the fundamentally new forms of the high-pressure or turbine locomotives – exclusively in finding ways to increase efficiency and load capacity. The aims of locomotive design in the immediate future should not be to seek for new forms, but to advance the present desire for expression to the limits of the possible.

From 'Gedanken über die Formgebung in Lokomotivbau' [Thoughts on the Shape of Trains], in Die Form *(Berlin), Jahrgang 2, 1927, pp 227-235, translated from the German by Roger and Agnès Cardinal.*

Philip Morton Shand 1930

Resistance to Rationalisation

Philip Morton Shand lived from 1888 to 1966. He was one of a handful of English writers who, in the pages of the Architectural Review *and elsewhere, introduced modern art, architecture and design to a wide British public. As well as being a writer on art, Shand was a well known traveller, bon viveur and gourmet. His published works include* A Book of Wine *(1926),* The Architecture of Pleasure: Modern Theatres and Cinemas *(1930) as well as translations of Gropius'* The New Architecture and the Bauhaus *(1935) and Adolf Loos'* Essays *(1936).*

Of all contemporary English writers, Shand was perhaps the most widely travelled and his European perspective is apparent in almost all his work. This piece of writing was first published as a short article in the Werkbund journal, Die Form. *It is the response of an Englishman with progressive tastes to the lethargy he perceived in the British industry around him. He deplored its conservatism and isolationism. Although he was, perhaps, too ready to adopt the Werkbund policy of* Typisierung, *that is standardisation and mass production, his criticisms will evoke a feeling of déjà vu in many readers today. That there was – and is – much to achieve in British industrial design is indicated by the relative poverty of exemplary British articles here in* In Good Shape.

One of the classical jokes in English literature occurs in Sir Thomas Browne's *Religio Medici*. The list of chapters of this ponderous volume contains one entitled 'On Snakes in Iceland'. Reference to this particular chapter discloses that it consists of the statement: 'There are no snakes in Iceland.' The present position in Great Britain in regard to the evolution of rationalised or simplified type-forms for manufactured articles in daily use is nearly as completely negative. This vital problem is not as yet understood by the public at large. There is an almost entire ignorance among manufacturers of the fact that it exists, and that on its progressive solutions will depend the future trend of industrial development. One of the difficulties in educating opinion in this direction is that there is no readily comprehended, or convenient, expression in English for what is known as *Typenwaren* in German. The technical vocabulary of Modernism is still in a too primitive and embryonic state for comprehension of the term to be generally grasped.

It is a melancholy and humiliating confession for an Englishman to make that the great movement towards standardisation of design in terms of functional fitness, to which the genius of the German people is now applying itself, should find no echo in the country which initiated the Industrial Age. But it has to be candidly admitted that no nation has more stubbornly resisted rationalisation in industry: To none – not even the French – are standardisation and mass production more abhorrent. This will probably seem all the more paradoxical to foreigners because Englishmen and English women usually seem to them the most standardised of mortals in appearance and opinions. In reality, under deceptive externals, no nation is further removed from approximation to the concept of *Massenmensch*. The depressed state of British industry and the progressive decline of our foreign markets are the direct result of innate conservatism, of a refusal to live and work in the spirit of the present age. The Briton is temperamentally incapable of understanding the significance of the word 'international'. Asked to define it, he would probably answer: 'The sort of thing they do on the Continent', which, he would leave it to be understood, he was thankful to say was not done in his own island. That our manufacturers are incapable of realising this does not alter the truth of the contention. British manufacturers are obstinately trying to go on living and manufacturing in the world of the nineteenth century. Our insularity, which is even more a mental than a geographic condition, helps them to continue in this fool's paradise. There is something lamentably 'amateurish' about the human atmosphere of nine British factories out of ten which is at once discernible to anyone returning from a business visit to Continental works engaged in manufacturing the same kind of goods. This atmosphere is exhaled by managing directors and technical staffs far more than by workmen and foremen. The former are not sufficiently educated in the nature and potentialities of the raw materials and processes which they employ. Too often they are either wholly ignorant of, or indifferent to, what is being done by their immediate Continental rivals. In matters of taste or new departures in design, they are inclined to be disdainful or suspicious of anything 'foreign'. One of the indirect results of this is that a 'modern' design implies something of an art nouveau type of decoration to the majority of British manufacturers and middlemen. Both have lost the adventurous spirit. They shun experiments. They will not look ahead. They have little or no conscious pride, or feeling of exhilaration, in the age they live in. Most of them, indeed, are so purblind as not to have realised that it differs in any material respect from the preceding one. They lure themselves into a feeling of false security, aided by the inherent British mental indolence, by believing that the twentieth century is simply a continuation of the nineteenth, and that the values of the two are largely interchangeable. Thus it is scarcely surprising that in Great Britain we can as yet hardly make any showing in the production of simplified *Typenwaren* if only because we do not even know what they are.

The English mentality in regard to machinery and machine-made articles is a difficult one to understand. The national pride in having once led the world in these respects is purely superficial. Beneath it lurks a belief, as passionately sincere as it is illogical, that there can be no question as to the superiority of hand-made articles. The machine is mistrusted as a bad and slipshod workman. The nation as a whole has yet to be converted to an appreciation of the fact that machine-made goods are more exactly made. The machine so completely conquered man in the England of the last century that it is hard to make modern Englishmen believe that in other countries man has already begun to conquer the machine. English manufacturers of the Victorian Age had seldom any real pride of craftsmanship in the goods they produced. More often their attitude was purely commercial. 'Quality' – the word as much as the thing – is probably a greater fetish in England than in any other country. Nowhere else do people of all classes and degrees of wealth respond quite so blindly and indiscriminately to shopkeepers' or manufacturers' appeals to purchase 'something rather better'. To buy the cheapest quality of anything – and this

Morris Oxford 6 saloon, 1930 (National Motor Museum)

quite apart from such considerations as its relative serviceableness – gives the Englishman an acute inferiority complex against which his pride involuntarily reacts. To his credit it must be said that few nations have a deeper respect for good workmanship and good materials. Rooted in the Englishman's mind is the essentially nineteenth-century idea, fostered by Morris and Ruskin, that machine-made articles must inevitably be inferior to, because a substitute for, hand-made ones. He does not see that they are simply different. In his scorn of *Ersatz* he refuses to realise that whole classes of goods have come into existence recently which it would be almost physically impossible to make by hand, and in which hand finish could not hope to be so satisfactory as machine finish. Wireless sets of British manufacture are, I believe, excellent in a technical sense. But the the cases of hardly any of them escape from some 'period' influence or other. The Englishman is not sufficiently logical to see that an 'Adam style' gramophone cabinet is a ludicrous anachronism: or that things so wholly new as wireless sets and gramophones should have cases designed simply and solely to enclose their mechanism in the most appropriate manner without any 'artistic' trimmings.

The civilisation of England is too old and has been too smoothly continuous to enable the average Englishman to be reconciled to the idea, which animates the American, that what he buys need not be of too permanent or durable a nature, because very soon something better will be perfected with which it will be to his advantage to supersede it. Such an attitude profoundly shocks the traditionalism of the English mind. The Englishman wants things that last a lifetime, or can be considered as possible heirlooms; the house he builds he wants to stand for all time. To secure this quality of durability he is ready to make the most disproportionate economic sacrifices, and does not understand that he is behaving as a 'defeatist' to the cause of human progress in consequence.

All the same there are certain hopeful signs. Men's clothes provide a good example, if only because their present international lineaments, both for everyday and sports wear, have been evolved and refined by an essentially English process of simplification and sobriety known as 'good taste'. Whether we find these clothes satisfactory, or even comfortable, is not the point. The point is that they represent a typical example of English 'whittling down' and reduction in design which in this particular case has now become universal. Sometimes this amounts to a refinement. Before the war it was an unheard-of thing for any educated Englishman to buy ready-made clothes. For it to have been known would have entailed a sort of minor social ostracism. This attitude was utterly illogical, and could only be explained by the notoriously bad quality of ready-made clothing because hardly anyone had hats, or shirts, or ties made to order, and people were quite content to buy these ready made. Yet since the war American methods of wholesale tailoring have been applied with phenomenal success. Most Englishmen are already convinced that they can be almost as well, and certainly as 'quietly', dressed in 'ready-for-service' suits at six guineas as in bespoke ones at 16. Indeed, it now needs a very expert eye to tell the difference between the one and the other. This is a first step in standardisation, and most significant. Since all classes and nations dress alike on general lines the Englishman's morbid dread of being sartorially 'conspicuous' is dying a natural death. Now that he has seen that the once despised ready-made 'reach-me-down' has been perfected by modern mass-production organisation to a high enough pitch of technical excellence for him to be able to wear it without qualms, it will not be very long before he begins to realise that articles in daily use such as furniture, china, glass and metal goods will gain equally by being standardised; and that by hastening the process of standardisation he will be accelerating a return to good workmanship and good design: things which are very near to his heart. But before this is possible he will have to accustom himself to the idea that hand-made and machine-made goods are essentially different in their nature; and that the era of unbridled individualism, with its concomitant hand-made goods, has gone for ever.

In pottery and glass things could hardly be in a worse state than they are at present. The great Staffordshire china-making firms are still resolutely living in the past and ignoring the present, to say nothing of the future. For the most part they are bent on faithfully reproducing the same shapes and decorative designs for which they were justly famous at the end of the eighteenth century, and at the beginning of the nineteenth. The full grotesqueness of the situation may be envisaged when it is stated that Messrs Josiah Wedgwood and Sons of Etruria have recently appointed as their chief art director an elderly gentleman, well known for critical monographs on the Italian Primitives, who was till lately the chief curator of the National Gallery in London!

Our general engineering products, where there is fortunately no scope for any 'artistic finish', have generally a simplicity and compactness of appearance excelling those of other countries. Our engineering draughtsmen are expert in tucking away all external 'gadgets'. English locomotives have still the cleanest and simplest lines – lines that are often of the greatest beauty and purity – of those of any nation. The English mass-production motor car, on the other hand, though as good as, if not better than, the French, German or American mass-production cheap car as regards finish, workmanship and durability, has about the ugliest lines of any. Once we turn to higher-priced, non-mass-production cars, however, we begin to find grace of line as in the Rolls-Royce, Lanchester or Bentley. Notable progress has been made during the last few years in the standardisation of materials and specifications by official bodies connected with the engineering, building and allied trades. But though these standards, when applied to things like the dimensions of sheet metals, are not without a certain influence on the size of the articles made from them, attempts to standardise shapes have hitherto been confined to things like steel rails, girders, fire-bricks and rolling-stock parts . . .

From 'Type Forms in Great Britain', in Die Form *(Berlin), Jahrgang 5, 1930, pp 312-314.*

Wedgwood Jasper ware, 1930s (Wedgwood Museum)

Coach-built Rolls-Royce Phantom II Torpedo motor car, 1929-30 (National Motor Museum)

London underground railway sign, Osterley station, 1934 (London Transport Executive)

Norman Bel Geddes 1934

Design is a Matter of Thinking

Norman Bel Geddes was born in 1893 and began his career as an advertising draughtsman and occasional playwright. He began to design industrial products in 1928 and with Raymond Loewy, Walter Dorwin Teague and Henry Dreyfuss he became a leader in the uniquely American field of product design. His theatrical background lent an air of showmanship to all that he did and his book Horizons (1934), *from which this extract is taken, has a spectacular and prophetic character.*

In his chapter on product design, Normal Bel Geddes still refers to the influence of machines over art, 20 years after writers associated with the Deutscher Werkbund had made this idea well known. The ideological innovation which writers like Bel Geddes brought into the design debate was the idea that good design had a commercial, as well as an aesthetic character. Bel Geddes and others managed to convince the American manufacturers they worked for that good design would pay dividends.

In this discussion of product design, Bel Geddes suggests a liberal interpretation of functionalism: he insists that mass-produced objects must be beautiful. He also insists that public preferences must be respected, although he was happy to try to convert public taste so that it would accept metal furniture, such as that which he designed for the Simmons Company in 1928.

The artist's interest in machines had laid the foundation for a new department in industry, in which the relations of product manufacturers and of consumers reach a new level of understanding and congeniality. The artist's contribution touches upon that most important of all phases entering into selling – the psychological. He appeals to the consumer's vanity and plays upon his imagination, and gives him something he does not tire of.

The designer of industrial products can only be successful if he is imbued with the conviction that machines, such as typewriters, automobiles, weighing scales, railway trains, electric fans, radiators, ships, stoves, radios, are good to look at when the problems involved are properly solved. An original creative aptitude for materialising this conviction in steel, wood, glass, aluminium, plastic substances and other materials old and new is the *sine qua non* of the profession.

A good illustration of the proper relation between use and appearance is the suspension bridge. A properly designed suspension bridge, regardless of its size, has the utmost simplicity. Its main supporting elements, the cables, hang between their supports as naturally and as gracefully as loose rope. The roadway is suspended from the cables by regularly spaced hangers. The location and direction of the cables and hangers conform to the natural lines of the action of the stresses within them, permitting the most economical use of material in their structional design. There are no superfluous or inefficient members. Inevitably, when all the elements of which the bridge is composed are organically assembled, the structure assumes a pleasing form.

There is an old saying that when a thing is

designed right, it looks right. In this connection there arises at once the difficulty that is usually involved in establishing a definition. The terms used may mean entirely different things to different persons. Picasso, for instance, could well make the same statement with regard to one of his compositions; and although Picasso might agree with the work of the engineer, the engineer is not likely to agree with the work of Picasso. An object is well designed when it has been reduced to its utmost simplification in terms of function and form.

While function, once arrived at, is fixed, its expression in form may vary endlessly under individual inflection. Form, referring to exterior appearances, always implies a high degree of quality, distinctiveness, and unity with its function. The public generally is unaware that a designer or engineer has enormous latitude in solving a particular problem in the right way. In this respect, engineering, architecture, painting, sculpture, poetry, music and all other media of design resemble one another. This semblance is the starting point of the trouble and it is this that makes the problems interesting. The correct solution of a problem depends on whether the designer is an artist or just a craftsman.

The first scale of the Toledo Scale Company, designed in 1897 by Allen de Vilbiss, functioned satisfactorily. In engineering terms, it was designed right and it looked right. Through years of use and experimentation the original scale was made to function better and also to look better. Its success made it the most popular and most widely used and imitated counter scale on the market. The problem of improving the design of this scale was given to me. My recommendations were slight and yet somewhat radical. There are two outstanding faults in the design of the existing scale: one its

weight, since it is made almost entirely of cast iron, and the other its large bulk. I redesigned the body to be made out of thin pressed metal, recommending aluminium. So that the purchaser may see simultaneously what is being weighed and its correct weight, the pendulum mechanism is located at one side and the cylinder mechanism cantilevered. The form of the enclosing body is simplified as much as possible and the only further recommendation I made was to set the scale into the counter, so that the weighing platform would be flush with the wrapping surface . . .

Much might be said of the necessary routine procedure in the creation of a new design before the designer so much as attempts a sketch. For the sake of concreteness, I will describe the routine that is observed in my own office.

When a new problem comes up for design, all my associates, those who will be connected with the problem in any responsible way, gather in my office and discuss the problem in all its phases. Depending upon the nature of the subject, different types of designers, engineers, merchandisers and research specialists are present. To ensure clarity of purpose and thoroughness, we proceed in accordance with a check list which has become more and more standardised with each new job. We determine the specific objectives and lay out the various means of achieving those objectives. A working schedule is laid out in weekly units. Later on, a detailed, day-to-day schedule covering all details of the work is made. These schedules are agreed to whole-heartedly by everyone in any way connected with them, and each person assumes responsibility for his or her part of the schedule. The discipline of maintaining a pre-arranged plan, schedule and set of restrictions is of great value. Continual analysis imposes integrity and directness

Mid-Hudson suspension bridge, 1925

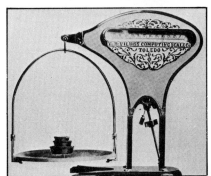

Toledo scale, designed by Allen de Vilbiss, 1897

(Photos University of Texas and Edith Lutyens Bel Geddes, executrix of the Norman Bel Geddes Estate)

Title page of Magic Motorways *by Norman Bel Geddes, Random House, 1940 (Photo Design Council)*

Airliner project no 4, designed by Norman Bel Geddes and Otto Koller, 1929

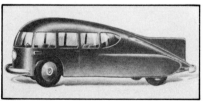

Motor car project no 8, designed by Norman Bel Geddes, 1931

(Photos University of Texas and Edith Lutyens Bel Geddes, executrix of the Norman Bel Geddes Estate)

in the mental processes and eliminates guessing and whimsies.

Our groundwork is founded wholly on facts. All records in my office are kept in writing. Verbal understandings do not count. All decisions arrived at in every meeting of consequence among ourselves or between clients and ourselves are covered by a stenographic record in the form of minutes. A copy of the minutes is sent to everyone present at the meeting, within 24 hours after the meeting, for approval or correction. This practice prevents misunderstandings and in several instances has prevented losses of thousands of dollars . . .

One great misconception is prevalent regarding design. Design is not primarily a matter of drawing but a matter of thinking. Personally, I do comparatively little drawing. The bulk of this is in the nature of preliminary sketches and in criticising designs in course of development. This is a matter of daily procedure. Every drawing, at every stage of the work, from the preliminary sketches up to the final shop drawings, passes across my desk for approval or criticism before it advances to the next stage. While drawings are being worked on, I go over them daily in the drafting room.

After a design has been approved by the client in finished sketch form, it is developed into the working drawing stage. By this I mean that it is restudied at larger scale which requires greater accuracy and a thorough consideration of details. Full-sized models are then made of wood or metal, depending on circumstances. They are completely finished, so that the client can get an accurate impression of the appearance of his product in its final form . . .

From 'Product Design as Approached', Chapter 11, Horizons John Lane, The Bodley Head Ltd (printed USA), 1934, pp 222-241. Reprinted with the permission of The Bodley Head and Little, Brown and Company.

Council for Art and Industry 1937

Art and the Technician

The Council for Art and Industry's 1937 report on 'Design and the Designer in Industry' was one of a number of official reports on art and design to appear during the 1930s, in part an official response to the activities of the DIA. At the same time, Nikolaus Pevsner published his Investigation into Industrial Art in Britain, *following research carried out at the behest of Birmingham University in 1934 and 1935.*

The Council's report was an important and far-sighted one, but everywhere in it the presence of lethargic British conservatism is betrayed. There is a persistent hesitation about investing completely in the possibilities of the machine and the Council appears to be anxious that 'traditional' hand craftsmanship should not die out in favour of industrialised automation. It is important to recall that at precisely the same time as this report was encouraging the employment of traditional values in industrial design, the Americans were adapting their manufacturing industries to the high-technology designs of a group of artist-engineers (like Raymond Loewy and Henry Dreyfuss) who were familiar with mass production and the commercial techniques and necessities of big business.

Nevertheless, the report represents an imaginative attempt to codify design practice for British industry, bemused by its failure during the twentieth century. Reading between the lines reveals everywhere a lackadaisical attitude to industrial design, but the conclusion looks to the future and argues for the designer to be accorded similar status to the engineer in the manufacturing process. Sadly, this position has rarely been reached even today.

The Place of Design in Industry

Definition of Design

At the outset of our Report it is necessary to offer some definition of the word 'design'. In its broadest sense design means planning in relation

a to function: the article produced must be fit for its purpose;

b to form, or, more widely, to aesthetic value with due regard to form, texture, colour and the aptness of any decoration.

Art is popularly supposed to be concerned primarily with the second of these elements, but the two are indissolubly linked with one another as well as with considerations relating to the selection of suitable materials and the processes of manufacture. There is a further essential element in design, of which the designer must not lose sight for purposes of production, namely, the economic factor – the necessity for producing an article at a price and of a kind which will command a ready sale in the market for which it is intended. All these elements in design, where manufacture is concerned, are so closely related that they form, in effect, an organic whole.

For many products, design is almost entirely functional, in the sense that either the purposes for which the product is to be used or the technical requirements of the process of manufacture largely control the design. In these cases design is perhaps more akin to engineering draughtsmanship. Precision and economy are its characteristics, even though here appearance has a definite value.

Even where form and decoration have greater play, there is still a natural relation between design and the material itself. The design should grow out of the material and be consonant with its qualities and characteristics . . .

In stressing the importance of craftsmanship . . . as the essential source of design, we are not overlooking the fact that our inquiry is mainly concerned with manufacture, the large-scale production of goods by machinery. The machine produces qualities of its own which have a value in themselves, and it should not attempt to copy handwork, though it is surprising how large still, in many industries, is the share of skilled handwork remaining to be done. The craftsman and the skilled machinist must be associated together, to secure that between them design is properly interpreted in terms of the machine process. There is, indeed, a 'craftsmanship' of the machine which should govern design in manufacturing processes and which in many directions still awaits exploration and definition . . .

By utilising the resources of artist-craftsmen, artists and architects, it is possible to enlarge the scope of international trade and even to set a standard in certain directions for world production. These examples should be a stimulus to British industry. There are signs of some tendency to decentralise the influences that have largely controlled certain ranges of current design. The dominant position formerly held in matters of fashion by Paris and Vienna seems to some extent to be breaking down. The British manufacturer has recently established an independent reputation in the field of furnishing fabrics (comparable to the already established reputation of this country for men's clothing and many other lines of production well known abroad), while New York is challenging Paris and sending to Europe clothing and shoes of definitely American design. These are new departures and suggest that, with perseverance, it should be possible, if we develop and utilise to the full our national resources of design, to secure for this country a more conspicuous place as a creator of taste and fashion ranking high in international opinion, corresponding to that which she has long held in the field of craftsmanship generally and in certain special branches of design for which she is justly famous.

Fashion is a dominant factor in design, and the field of fashion is extending from short-lived articles, such as dress, to articles of longer duration of life. When articles are to be replaced at shorter intervals, there is naturally a demand that they should be cheaper, and in consequence a tendency to sacrifice durability and intrinsic quality to the necessity for a lower price. Whether this tendency will be permanent it is impossible to say; but meanwhile it cannot be disregarded, and the result is that fertility in design is now more necessary than before and a prolific output of good work from industrial art designers is of increased importance to industry . . .

It is necessary, too, to consider the various attitudes of these people [designer, maker, salesman, customer] to questions of design. At certain stages the influence of commercial considerations is predominant. The rate of turnover or the cost of marketing may determine the selection of designs. Economic considerations may easily exercise a preponderating influence upon merchanting, and the ultimate business management is often in the hands of men who are unversed in the considerations of taste that should enter into a choice between designs. There is an inevitable conservatism in the maker, born of market considerations, and an inevitable caution in the distributor who has to take the risk of trying to sell particular goods. New design, therefore, before it can reach the customer, has to run the gauntlet of criticism, and selection at many stages, dictated by considerations far removed from aesthetic values, and the various checks tend to retard its development except in so far as the aim is novelty for novelty's sake.

Sources of Design

The main sources of design in the industries which we have considered are:

a the design room attached to the factory;

b designers who visit the factory from time to time or are specially employed for varying periods;

c freelance designers who work outside industry and produce designs to sell to manufacturers.

. . . It is desirable that the design room at the factory should to a much greater extent become a centre of creative design and be less absorbed in mere adaptation, and that where the purchase of designs from outside is necessary every effort should be made to maintain some degree of contact between the internal design room and the external source from which the designs are obtained.

The shortage of original, creative design in some of our industries at the present moment can be illustrated by reference to certain industries which have recently been created. One or two of these industries are now developing a series of new materials easy to work and readily adaptable to a number of purposes. These materials, while possessing definite characteristics of their own, form effective substitutes for many older and more difficult materials. Almost anything can be done with them, but the new industries have not yet fully explored their inherent possibilities and limitations and are producing but little fresh design, related to the properties of the material itself. Instead their designers often tend to surface ornamentation to cover defects in workmanship and aim merely at copying other . . . materials and processes – tortoiseshell, ivory, glass, wood, marble and even marbling and graining – although the essential properties of the material make it possible to secure by natural use effects richer in colour and finer in form than those based on imitative processes. No real progress will be made on these lines.

Creative design for manufacturing purposes is chiefly derived in this country from the ranks of the external or freelance designers. In the fashion industries, there is a preference for design from abroad on the ground that fashions in women's dress are largely controlled from the Continent and that foreign designs are likely to be more original and to supply the touch of novelty which is so important as a sales factor. It appears, however, to be the fact that some of the designs purchased by our manufacturers as 'foreign' designs are really home designs which have been exported to secure a foreign cachet and an enhanced value in the home market.

It is, of course, necessary in all cases to bear in mind the work done abroad, if only as a guide to foreign markets and to maintain touch with the trend of international fashion. One of our witnesses remarked that 'a firm which cuts itself free from foreign ideas and inspiration will cease to be a producer of novelties'. We believe, however, that there is ample scope for . . . British freelance designing, and we hope that it will be possible to establish in London a market for British design comparable to the design markets found in certain Continental centres.

. . . In addition to the main sources of design so

far discussed, a further source is found among independent artist-craftsmen, who as we have mentioned have in the past had an important influence on design and can still be of the greatest value to industry. These craftsmen now supply a restricted circle of customers with articles which may later be adapted by industry for factory production. Sometimes they are commissioned to make sample pieces which industry acquires and multiplies. Industry will be wise to make a still fuller use of the designing ability of these craftsmen for industrial purposes.

Architects, painters and sculptors, working independently, more rarely contribute at present to industrial design, but they would do so to an increasing extent if they were patiently encouraged and if they accepted the discipline of industrial technique. Selected individuals from among them, as well as from among the artist-craftsmen, would contribute a freshness and originality of mind which would be of the greatest value to industry.

The Status of the Designer in Industry

Broadly speaking . . . at present industrial art designers have failed to attain a sufficiently definite and established place in industry. Too frequently their status is entirely subordinate, and in consequence they exercise little control over the output of industry. While the chemists, the engineers, and the technicians generally are highly organised, with professional institutes which attest their capacity and watch their interests, the designers have no really authoritative and comprehensive organisation of this kind to rely upon. Apart from the Society of Industrial Artists, a few guilds exist here and there, as in the Macclesfield Industry, or at Stoke in the Pottery Industry, but there is no fully developed body comparable to the established professional institute, carrying with it a recognised professional status and commanding general recognition, and as a result the interests of designers have suffered. A proper organisation should be created as soon as possible.

It is interesting to consider why designers are not accorded the same respect in industry generally as that freely given to the expert technician, and why the recent advance in science and technology has not been accompanied by a corresponding advance in industrial art. One point to remember is that technology has an obvious, direct bearing upon the methods and economy of production, and therefore appears to the manufacturer to be a much more vital factor in production than artistic design, the commercial value of which is not always understood. Moreover, science and technology permit of exact statement and evaluation, whereas artistic merit is to some extent a matter of taste and most people are not qualified to make difficult aesthetic judgments.

The average industrialist is often inclined to distrust as someone unpractical and difficult the fully trained art school student who seeks to become a designer in industry . . .

Something must be done to improve the position. Not only must the industrialist receive some training in design and extend his research habit of mind into the design side of his business, but the artist must also adjust himself to new conditions. With the invention of the more elaborate tools represented by machinery, the discovery of new processes and materials, and the development of mass production, technical considerations must more and more control and direct design, and it is therefore becoming of increasing importance that the designer should have a sufficient understanding of these considerations if he is to play his part and commend himself and his art to . . . industry.

From 'The Place of Design in Industry', Part II, Design and the Designer in Industry, *report by the Council for Art and Industry, HMSO, 1937, pp 7-15. Reprinted with the permission of the Controller of Her Majesty's Stationery Office.*

The Responsibility of the Public

Nikolaus Pevsner was born in Leipzig in 1903. He has done more to introduce modern design to the British public than any other writer of his generation. To the British scene he brought the advantages of the German critical tradition and the independence of mind which it engendered. Pevsner was considerably influenced by the propaganda of the Werkbund and by Bauhaus values and has maintained throughout his life a certain puritanism, tempered by a commitment to beauty. Both these influences contributed to his book, Pioneers of the Modern Movement *(1936) which, although unsuccessful when first published by Faber, achieved a dominance in its subsequent Pelican editions which tended to establish an historical tradition of modern design. The dominance which Pevsner's forceful and tireless arguments created have recently produced a reaction by revisionists who, temperamentally out of sympathy with the aesthetic and ideological aims of the Modern Movement, deplore Pevsner's determinism. Yet Pevsner himself has remained true to his early opinions.*

This extract is from Nikolaus Pevsner's An Enquiry into Industrial Art in England, *a study undertaken at the instigation of Professor Philip Sargant Florence of Birmingham University. It is among the first books that Pevsner produced after his arrival in England in 1933. On publication, it was generally well received, but did not immediately achieve great popularity. It is especially interesting because in the* Enquiry, *Pevsner applied to Britain, its industry, its society and its market, the ideas and principles which informed his historical writings on modern design.*

The Taste of the Public

It would be absurd to deny that there are marked differences between the taste of one group and another, one nation and another. I mentioned several times the experience of manufacturers as to contrasts between London and the Midlands, England and Scotland, Great Britain and the Continental countries, Europe and South America. I referred to the fact that Oxford is supposed to prefer Oriental to modern carpets, Scotland to prefer heavier and more practical leather handbags, a stronger and better quality of linoleum, and more ornate and solid-looking type of pottery. Generally speaking, most Continental countries go in for brighter and gayer things than England; London goes in for smarter, lighter and more delicate-looking things than the North (in furnishing fabrics; Manchester asks for heavy weight and heavy colour; London for rayon and pastel shades); and Edinburgh for brighter and more showy things than Glasgow. While within Britain a tendency is evident by which new fashions are developed in London and then gradually move into the Provinces and up to the North, it cannot be said in the same way that new artistic styles always come from the Continent into England. For one thing, to speak of the last centuries only, the Gothic Revival and the Morris Movement originated in England; and then there are Continental modes of expression to which England seems almost entirely immune. During the seventeenth and eighteenth centuries, hardly any excessive baroque or rococo forms were used in England, while they were predominant in Italy, Germany and, up to a point, in France also. When Paris from 1860, and Berlin from 1875, began to revel in neo-baroque ('Wilhelmian' style), in London Norman Shaw and his friends erected well and soberly designed houses in a modernised Queen Anne and Georgian style. This English sobriety seems unalterable, and in fact so essentially English that it would be a crime to alter it. Britain's greatness appears inseparable from Britain's conservatism . . . Experiments and ensuing failures are instinctively left to other nations. However, while this basic English quality accounts for a certain hesitation in the adoption of the Modern Movement, the modern style in its entirely cannot be one of those which England will in the long run refuse. For it is a simple and rational style, and moreover a style in the creation of which England has had a considerable share . . .

The most serious obstacle to its [the Modern Movement's] divulgation lies in the attitude of the upper class. Owing to the general conservatism mentioned, to inborn reserve and a distrust of anything that looks strikingly new, the majority of the English upper classes, above all the aristocracy, still prefer period decoration, period furniture, period procelain etc – whether genuine or reproduced – to modern industrial art. In fact, I had more than once to refer to a new increase of period reproduction which has been noticed by manufacturers and retailers within the last year. The reason for this was revealed by a headline in a recent issue of a furnishing trade paper: 'West End Trade buys Period Design. Modern schemes now too popular.' This smacks badly of the 'Something-that-is-different' attitude, but it may also mean that modernism in its jazz forms has spoilt the market for serious modern work.

If that is so the consequences will be disastrous. Snobbery – the wish to imitate, if not in fact, at least in outward appearance, an envied class of which you read in the Court and Society column of your morning paper – snobbery could be a great help to the growing Modern Movement in England, if only more members of the upper class would give up Chippendale for modern furniture of equally high craftsmanship and perfect design, and old Chelsea for Keith Murray. At present it appears to be the professional class mainly, and a small minority of wealthy merchants and industrialists who uphold the modern style. Still, their taste is also bound to filter down by degrees into the semi-detached houses of the poorer middle classes, and the question now is whether these are prepared to accept the new simplicity. Most manufacturers and distributors say that their predilection for the bad, the meretricious and the showy is ineradicable. Time and again I asked them how they could know . . . Manufacturers . . . depend largely on suggestions from their own sales departments. These are based on information received from the firms' travellers who come home and state what has been a 'flop' or a 'washout', and what is selling like 'hot cakes' or like 'a house on fire' or like 'ripe cherries', according to the traveller's favourite metaphor. But is this really an indication of the popular taste? The traveller in his turn relies on statements which he hears from the buyers of department stores or shops, and from shopkeepers. And as to the shopkeeper, he unconsciously pushes what he likes and conceals in dark corners what seems to him unattractive . . .

In these circumstances the manufacturer, supposing he is not an exponent of the same good-old-days mentality, ought to try to find out whether the information which he receives from the sales department is a true interpretation of the wishes of the public. He can do that easily in the few cases where a firm has shops of its own.

Otherwise he can attempt to check his travellers' reports by sometimes travelling himself, or by going to shops and enquiring privately, or by showing new models to as many friends and employees as possible, or by organised experiments of the type carried out by Messrs Cadbury and Messrs Foley China, or by trade questionnaires such as I mentioned in connection with radio firms, or by other questionnaires on a large scale. I quoted the results of my own experimental questionnaire, which sufficed to show that it would be worth some money and trouble to try something of the sort in a more expert way. Much too little has been done by British producers to obtain a more objective view of the popular taste. The results of any such enquiry would no doubt be surprising. We saw that in some trades, such as door furniture and cheap electric fittings, the wishes of the public hardly count. Almost all business is transacted with builders' merchants. In buying or renting a cheap house, it is only in exceptional cases that the client will express his view of the door handles, or of the wall brackets and pendants, which he sees in the rooms. Here then the popular taste – and, as I said, a distinctly hideous taste – is not the taste of the public but that of the builder and his merchant. One may say that this probably comes to the same, but it ought not to be forgotten that the same builders have recently shown a predilection for panel fires, ie a type of fire greatly improved in design, and for plain distempered walls (notwithstanding the fact that when they feel it their duty to be artistic they go in for excesses of appliqué wallpaper). The same type of wholesale merchant in the jewellery trade, though he insists on conventionally and thoughtlessly designed brooches and bracelets, will refuse wrist watches which are not simple and plain . . .

Ferranti model 3310 electric fire, 1937 (Design Council)

Burney Streamliner motor car, 1930 (National Motor Museum)

Edward Rumpler's saloon, 1921 (Deutsches Museum, Munich)

Pel RP6 nesting chair, 1931-32 (Design Council)

In some other trades the decisive importance of the manufacturer may emerge even more clearly. I referred to streamlined cars in that connection, also to steel furniture and to modern banded, as compared with modern floral, pottery. In each of these cases producers have deliberately pushed a certain line, pushed it consistently and over a sufficiently long time, with the result that it became the fashion and the quantities of sales expected could be secured. I say over a sufficiently long time, and that is an essential point. We heard that Messrs Ferranti had to wait some years before their new type of fires became popular; in banded pottery the time-lag between the start and the universal acceptance was about three years; in streamlining, between Sir Denistoun Burney's initiative and the conquest of the English market, about four years; in English tubular furniture about four years again. This shows how absurd and unpardonable a practice it is in some stores to give a new article one month's or three months' trial on the counters and then to drop it for ever if sales have not reached a certain level. To call this investigating the popular taste sounds like satire.

Considering all these conflicting facts and arguments, the only statement of the public taste which can be made is that 'the public does not know itself what it wants', whereas most distributors and manufacturers know pretty well what they want . . . Perhaps after a few more years of propaganda, perhaps after a generation, they would no longer relapse into certain atrocities. But under present conditions, their taste is a blank. They take what they are consistently offered. Ruskin has preached that the business of the manufacturer is to form the market as well as to supply it. He might have added that whether the manufacturer wants it or not he is forced to form the market, for better or for worse.

However, a later section will be devoted to the duties of the producer. In our present connection, it would not be fair to deny some obstacles to perfect design in the public itself. A certain admiration for elaborate craftsmanship of the same type as that found in the sixteenth-century crucifixions with 52 persons present, all carved in a cherry-stone; or of modern Canterbury Cathedrals made by enthusiastic amateurs from cork, or lard, or sugar, is a natural outcome of a simple, unsophisticated mind. So is a certain admiration for 'kitsch'. On the other hand, perfection of design cannot be welcome to most human beings, because it is too exacting to live up to. To see Raphael's Sistine Madonna, Michelangelo's ceiling, Rembrandt's Christ at Emmaus, Grünewald's Isenheim Altar once a year can be a sublime experience, but one would not like to live in the same room with any of them. However, allowing for human frailty is still no excuse for jagged door handles and vulgarly decorated cups and saucers.

It is agreed that the public could not buy these unless they were put on the market. But if people disliked them as thoroughly as one would wish them to, they would not buy them even though they were on the market. The fault of the public in the matter is that very few people take as much trouble in buying a lamp or a vase as they do in buying a hat or tie. Not that there are not vast numbers of people who really enjoy the atrocities which shopkeepers display. They are accomplices to the crime, no doubt. But while there is no justification for the educated and wealthy managing director of a large firm producing rubbish, there are several excuses for the humble consumer, excuses derived from the social conditions in which he is compelled to live.

The Social Problem

No exact figures are available as to what proportion of the English population grows up in slums and semi-slums. Official reports tend to be too lenient. An independent investigation carried out in Birmingham in 1933-34 showed that 28,000 houses should be cleared at once and 18,000 back-to-back houses as soon as possible. The word slum has become a common term for unbearable housing conditions all over the world; rickets, the disease mainly bred in slums, is called the English Disease in Germany. Not that overcrowding and its consequences are confined to Britain in any way. Things are equally bad in other countries, worse in some, better in a few where great building activity was possible immediately after the war. A recent questionnaire filled in by 900,000 schoolchildren in a country which is supposed to be exceptionally clean has brought out that 39.4 per cent of the children use no toothbrush and 18.2 per cent a family toothbrush. No wonder that in such conditions no sense of beauty and hardly a sense of tidiness can be developed. And no wonder either that the natural longing for beauty, inborn in almost everybody, slakes itself on the cinema, on showy furnishings and 'kitschy' pictures in the house and on the portraits of film stars – heroes and heroines of dreams of wish-fulfilment – in the workshops.

Design in daily life is not a detached question, this cannot be said too often. No conclusive improvements can be made in design unless social improvements are achieved before and alongside them. A sweeping change of social conditions, such as the establishment of some kind of State Socialism, might lead to a sweeping change in the appearance of industrial products, although it seems impossible to generalise on the aesthetic consequences of such a change. In Germany the post-war Labour Government fostered the modern style, whereas National Socialism tends towards the protection of handicraft as opposed to industrial design. In Italy modern architecture and modern design enjoy the special furtherance of the Government as being an expression of Fascism. In Russia the same style was, after some years of State approval, given up as an outcome of latter-day bourgeoisie, and a decidedly conventional and therefore probably more popular style launched. Whereas the reaction of art to changes within our civilisation thus remains indeterminate, the consequence of a breakdown of the whole of our civilisation and the beginning of a new circle is evident. The 'Merovingian Age' which would necessarily be the first phase to follow would at one blow get us back healthy though primitive conditions of craftsmanship, the unity of design and production, and Morris's 'joy in making' which is definitely lost as long as present conditions last. It is a matter of personal outlook and taste whether one regards the price to be paid for a true revival of art and craft as too high . . .

It remains doubtful how much of the desire for better design which we have noticed, and which is evidently growing, is based on a genuine appreciation of beauty, and how much on a craving for visible social distinction. Often it is bewildering to see that the same wealthy industrialist whose house is furnished in a pleasant taste, works in a dingy office which you reach by driving or walking through filthy streets, passing some black sheds with blind and broken windows and climbing up a dark, narrow and dangerously steep staircase. To explain this strange contradiction, it is necessary to refer to the deep-lying tradition which Puritanism has become in England. Work is our duty, but duty is something stern and forbidding that one ought not to embellish . . .

It has appeared to me advisable to call attention to many changes that are urgently needed, because occasional improvements of design cannot be of more than ephemeral value so long as conditions of life remain as they are. But it would be a grave error to leave industrial art alone for the time being in order to devote all available energies to other fields. The battle has to be fought on all fronts. Not one of the subjects is less essential, not one can be neglected, neither slum clearance nor the renovation of school buildings, neither the levelling up of class contrasts nor the raising of standards of design . . .

From An Enquiry into Industrial Art in England *Cambridge University Press, 1937, pp 204-215. Reprinted with the permission of the author.*

Anthony Bertram   1938

De Gustibus non est Disputandum

Anthony Bertram read English at Oxford and soon became one of the best known writers on art among a generation that included John Betjeman, James Richards and John Gloag. He was occasional art critic of The Spectator, *playwright, novelist, broadcaster and the author of a book about Rubens. For a man of such liberal education, his writings on architecture and design show just how far the ethic of the Modern Movement had penetrated the English intelligentsia by the late 1930s.*

Design, from which this extract is taken, is perhaps Bertram's best known book on art. It was one of the first Pelican Specials and with it Bertram caught the mood of the time. The source of the book was a series called 'Design in Everyday Things', broadcast on BBC radio during 1937. At a time when British institutions such as London Transport and the BBC, most especially through its journal, The Listener, *were investing spiritually and financially in modern design, Bertram's persuasive popularisation of austere Continental ideology made his text a classic of its type.*

In 1869 Charles Eastlake, an architect and designer, published a book called *Hints on Household Taste*, which he claimed to be the first publication on design 'in a manner sufficiently practical and familiar to ensure the attention of the general public'. 'To ensure' . . . how optimistic those Victorian reformers were. Since then there has been a crescendo of such books. Design is the subject of university lectures and of periodicals in all languages. It has been the subject of many broadcasts. There is a Government Council for Art and Industry, a National Register of Industrial Art Designers, and there are private organisations such as the Design and Industries Association devoted to propaganda for good design. And yet most of the abuses Eastlake complained of survive and the very word 'design' is a mystery to the common man, almost a clique-word. As it is used today – in the title of this book, for example – it implies a whole cluster of things connected with an object: its purpose and the plan of it, the object itself, its quality, material, usefulness and beauty; even the price and method of manufacture of it. Moreover it is a relative word. Though there are general principles of design, we can only judge whether any particular object is well or badly designed in relation to the particular problem it tries to solve. This is merely rather a pompous way of saying that a chair must be judged as a chair, and what is more as a wooden dining room chair at a low price for a working-class kitchen or an upholstered easy chair at a high price for a luxury drawing room or a metal chair at a moderate price for an office, and so on. It is not enough to say that a chair solves its problems if it can be sat on. Whether it is well or badly designed depends on what it is made of, who is to buy it, who sit on it, for how long and for what purpose: and on what sort of looks it has.

By 1588 the word 'design' had the meaning 'purpose, aim, intention': by 1657 the meaning 'the thing aimed at'. In 1938 it has gained the composite meaning of aim plus thing aimed at. It has come to stand for a process – from the original conception through the plan and the manufacture to the finished object.

By design, then, we do not mean a drawing made in his studio by the designer – a design *for* something – but rather the thing itself, but with that drawing, the preceding idea of the thing and the succeeding process of manufacture all implicit . . .

The idea of honesty is fundamental to good design. A well designed object should not only serve its purpose well but should look as if it were made for that purpose and no other. For example, an electric grandfather clock which houses a cocktail cabinet in its case does not suggest cocktails, but weights and a pendulum. Similarly, stained deal laths tacked on to the front of a building suggest that it is a timber-frame building. It isn't. But this dishonesty is not about purpose but about method of construction. Or again, paper painted to imitate marble or the grain of oak and pasted on to white wood is dishonest about material. So that we may say that a well designed object must be honest in three ways. It must confess the purpose for which it is constructed, the method by which it is constructed and the material of which it is constructed. But these three elements of honesty in design are not really independent. The purpose will usually suggest and sometimes dictate the method of construction and material. The material will often dictate the method of construction, and so on.

So it might seem that to judge the design of an object is a more or less mechanical process, for which we only need a little common sense to know if it is fit for its purpose, and a little technical knowledge to know if it is honestly made. But is this really how we judge? Do we not, in fact, in the case of all not purely utilitarian objects, choose by looks? Ask any furniture dealer or house agent. He will tell you that it is looks that sell. There is no space to discuss here why this is so. It has been true ever since man began to make things. And here we are up against the real difficulty. Very few of us agree as to what good looks are: or whether any given object has them or is a work of art. The most popular of all Latin tags is *de gustibus non est disputandum* – you can't argue about taste. But can't you? Is there any subject on earth people argue about more? And when they have finished up by being really rude to one another, they mutter *de gustibus* and go away convinced that they are right . . .

I will brave the anger of the untrained and state that it is simply not true that everyone is born with the capacity to judge design. As with every other capacity, even the greatest 'natural gift' must be trained. Without some knowledge and directed practice, it is no more likely that a man would make an accurate judgment in this matter, than he would in law or medicine.

Are there, then, rules of taste, standards of beauty, tests of art? To some extent, yes. At least there are guiding principles. At least certain signposts and danger signals can be set up, certain blind roads indicated . . .

Art involves the idea of human creation, of a thing made by man. There is no reason on earth why it should look even remotely like a natural object. A textile that merely looks as though a lot of daisies had been sprinkled on it, is most emphatically not a work of art, however skilful a work of imitation it may be. And just as there is no art in imitating nature, so there is no art in imitating another work of art. A good design must always be original.

That does not mean that it need be freakish. Unless there is a reason for a new design, it is better to stick to the old one . . . A well designed teapot is not any old teapot 'beautified', but a new teapot that is made beautifully. This distinction is not a mere quibble over words. There are two very important ideas in it. First, that the intention to produce a work of art should be present right through the design of the object, right through the process, that is, from conception to finish. Secondly, it means that ornament is not *necessary*, but it does not mean that ornament should *never* be used. Ornament is always an addition, and whether there should be any on a particularly object, and if so what, is a separate problem that we must return to. But first of all the essential thing – what does 'making beautifully' mean? It means that the shape, colour and texture of each part of the object should be as pleasing to the eye as possible without interfering with its efficiency, making unsuitable suggestions or pretending to be what it isn't: and that each such part should be related to the whole and to all other parts, as the parts of a sonata are related to make one piece of music.

This may sound rather pretentious when we may be thinking of a sewing machine or a kettle or a cheap bungalow. But not a bit of it. Art is completely democratic. It demands as high a standard of beautiful making for a bungalow as for a palace: as high, but different. That is what is implied above by 'unsuitable suggestions'. The sewing machine that is decorated with gold transfers of ornament borrowed from Gothic altar-pieces and stands on cast-iron imitations of rustic work is making unsuitable suggestions, dressing itself out in a tawdry finery and a bogus rusticity which have nothing in the wide world to do with sewing machines. The average bungalow today wavers in appearance between the 'bijou baronial' and the 'Tudoristic': that is to say, it makes an exceedingly bad shot at looking like a stone castle built for the wicked uncle in the pantomime or like a primitive timber and daub dwelling built for a mediaeval agriculturist. If the owner were logical he would wear cheap tin armour or hodden grey. In fact he wears a cap and a reach-me-down and maybe a bowler hat and a gent's suiting on Sundays, because he is neither a knight or a villein, but Mr Smith of 'Osocosy'. What odd frustrated dream is it that makes him put his home into such makeshift fancy dress . . . ?

All ornament of the past was evolved from processes of hand making. It grew naturally out of the tool and the material, and the creative urge of the man using them. All this is changed. The majority of goods today are machine made, or assembled from machine-made parts; and the artist-craftsman is a rare and expensive person. The result is that in most cases where ornament is applied, it is made by a machine to imitate a hand process; and very badly it does it. Another piece of dishonesty. Look at the stamped ornament on cheap furniture. Only a very crude eye could mistake it for carving, but that is what its shapes feebly imitate. It has no new shapes that have grown out of the new process. It is therefore unsuitable in another sense – unsuitable to the method of construction.

For these two reasons, it will be seen that good ornament is only possible today on relatively expensive goods, the sort of things we use in our

Kettle designed by Lurelle Guild, 1934 (Martin Greif)

Murphy television set cabinet in rosewood, pearwood and sycamore, designed by R D (Dick) Russell (Design by Anthony Bertram)

Shell poster by Graham Sutherland (Shell UK Oil)

Project for a slab block of flats, designed by W A Eden, 1936 (Design by Anthony Bertram)

best clothes; though, in some cases – pottery and textiles, for example – the cost of good ornament is not excessive. But the important matter is the shape of the object itself – its form. And that, like the forms of men and animals, can be quite sufficiently beautiful without trappings and trinkets, make-up and tattooing . . .

We need a special body of men trained in aesthetics and having a sufficient knowledge of technical processes, familiar with the history of visual art as a whole, and capable of imparting their knowledge and ideas. In short, we need trained teachers in the appreciation of design.

And who is to be educated? Everybody of course: but in particular the children. In every school there should be the ceaseless example of well designed surroundings, regular lessons by the experts – it should be part of an art master's job to be such an expert – practical illustrations of good and bad design on exhibition, and so forth. Among the adults, those whose education would most rapidly improve matters are probably the shop-keepers, buyers, salesmen and commercial travellers. Heaven knows, the manufacturers and designers are not all perfect; but there are plenty of good designers and many wise manufacturers, but it is difficult for them to get their goods to the public through the barrier of untrained and timidly conservative retailers – among whom, of course, there are honourable exceptions.

And the public, the buyers of things? What about them? They must choose a house and innumerable objects to go into it: they must use many buildings – town halls, swimming baths, churches, cinemas, pubs . . . All their lives they are inescapably surrounded by design. The other arts they can escape; but not this. They need not read, look at pictures, hear music, go to the theatre: but they cannot avoid design. And yet it is precisely that art which is usually quite neglected in their training for life. The public can learn to create a demand for well designed things. But they can only learn by seeing and hearing of well designed things; and if the shops do not display them, how are they to do this . . . ?

This is only the beginning of the story. We must have good architects in all our cities and rural districts to control public building. We must have laws to give . . . architects control also over the shameful building by irresponsible private speculators that is ruining our country. We must kill by ridicule the absurd 'ye olde worlde' cult that has infected English design with dishonesty. And, above all, we must have the widest propaganda against the idea that good design is a matter of expensive luxury stuff following a swift fashion.

Good design is not a matter of wealth, much less of the chic, the latest thing. It is not a matter of novelty for the sake of novelty, but of the production of cities and houses and goods which will best satisfy the needs of the people; their need of practical, honest, cheap, lasting and beautiful things to use and see in their everyday lives.

From 'What is Design?', Chapter 1, Design Penguin Books, London, 1938, pp 11-19. Reprinted by permission of A D Peters & Co Ltd.

Garage or Temple?

Walter Dorwin Teague lived from 1883 to 1960 and was one of a handful of industrial and product designers who shaped the appearance of the modern world. Two jobs, his corporate identity programme for Texaco filling stations and his interior design of the Boeing 707 jetliner, are perhaps his best known contributions to popular taste.

This extract is a part of a chapter called 'Fitness to Function', published in Teague's book, Design this Day *(1940). The year 1940 was something of an* annus mirabilis *for publications by leading American industrial designers: besides Teague, Harold Van Doren published his book,* Industrial Design, *and the second (English) edition of Normal Bel Geddes's* Horizons *also appeared.*

Teague offers his readers an optimistic view of the consequences of machine production, believing, as he did, that whatever works well must inevitably look good too. This notion, derived from Louis Sullivan's architectural theory via Frank Lloyd Wright, is insisted on throughout the book. Complex, high-technology machines such as aeroplanes are especially admired by Teague because they provide unique opportunities for novelty and experiment, as they have no residue of traditional form to hinder the designer.

Speaking admiringly of increasing speed and the potential of mass production, Teague was an unashamed advocate of the technocratic, consumer society. In support of his own aesthetic preferences Teague refers time and again to the argument that the most beautiful forms, as in streamlined aeroplanes, are created by necessity.

This text is one of the definitive statements about industrial design by an industrial designer.

. . . The Victorian era had reason to blush at utility, for most of the utilities offered it by an adolescent Machine Age were indecently hideous. It was easy for Morris and his converts to set these crude mechanical products in humiliating contrast against the suave works of the old handicrafts, not stopping to think that the craft forms stood at the end of a long process of evolution while the machine forms stood at the beginning of another. The perfect adaptation of the older products to the uses for which they were intended, their simple, honest effectiveness and the deep-rooted, organic source of their beauty were serenely evident, and Morris's campaign had a brief prosperity. It could not succeed because it was a mere eddy in an irresistible current, a last revolt against destiny.

The object of work is not only its own joy. While there should and must be joy in work, its object also is the increase of material aids to the comfort and richness of life. Machines are capable of making more things and better things, making them faster and cheaper and available to more of us, than is possible through handwork alone. The end that Morris held most dear, the communal happiness of mankind, required that machine production should prevail, as a later generation of social revolutionists in Russia clearly realises. But Morris did a great constructive service: he was one of the first to revive appreciation of sound and honest work and to re-establish rightness as the aim of all our mechanical endeavours. He made utility respectable and demonstrated the compatibility of utility and beauty. He maintained with lusty force the venality of ugliness and he convinced a considerable portion of an astonished public that a frame of satisfying beauty for the normal human

*Douglas DC-3 aircraft, 1934
(McDonnell Douglas Corporation)*

*American Sales Book Company Wiz
Register, designed by Walter Dorwin
Teague, 1934 (Walter Dorwin Teague
Associates)*

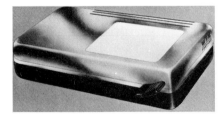

life is not only possible but actually essential to our self-respect . . .

After gaining momentum slowly for the better part of a generation, this new craftsmanship of the Machine Age is practised now by most of the alert and young-minded workers in Europe and America – architects, engineers, designers – who are attempting to devise forms to fit the functional needs of our times more exactly than any we inherited from the past. In all the fields of communal interest – cities, building, housing, transportation, manufacturing and public works – this modern school is practising a craftsmanship aimed at the creation of order by means of our own special equipment to meet our own special needs. It is attaining the ideal of honest workmanship at which Morris aimed, but by the very means that he thought hopelessly evil.

For the spirit of the craftsmanship is independent of its tools. It can work with steel presses and milling machines as well as with hand planes and chisels, if its aims are honest and its intelligence adequate. As modern engineering has advanced in mastery, designing for purely functional ends alone, it has created examples of perfected order that meet all the high standards of Sullivan and Wright, and it has done it often without ever having heard of these gentlemen. In the superlative rightness of certain modern airplanes, power plants and machine tools, parkways and bridges, nothing has been admitted which did not contribute to performance, and forms have been determined solely by efficiency, materials and processes; while an accurate integration of all the parts in precise relationships has been achieved by the pressure of necessity. As a result these things approach a classical, abstract beauty of form which advances toward perfection . . .

This is inevitably true, that as a thing becomes perfectly adapted to the purpose for which it is made, and so approaches its ultimate form, it also advances in that power to please us which we call beauty. Use is the primary source of form. The function of a thing is its reason for existence, its justification and its end, by which all its possible variations may be tested and accepted or rejected. It is a sort of life-urge thrusting through a thing and determining its developments. It is only by realising its destiny, and revealing that destiny with candour and exactness, that a thing acquires significance and validity of form. This means much more than utility, or even efficiency: it means the kind of perfected order we find in natural organisms, bound together in such precise rhythms that no part can be changed without wounding the whole . . .

So it is with all design in nature, and in the things we make. It is *rightness* that gives us pleasure, entices and thrills and satisfies us, and rightness is a revelation of the function for which a thing is right; a revelation too of the skill and soundness with which it is made, and its adequacy to perform its function. When we find all these factors of fitness evident in a thing we have made, it delights us with the complacent knowledge that we have added something to the total of humanised order in the world.

The function of a thing may be trivial or magnificient; it may serve a very humble or a very exalted end. It may be a model of a four-masted schooner inside a bottle, serving no other purpose than to display the maker's skill and tickle our fancy; or it may be a 40,000-ton liner, carrying thousands of people swiftly and safely and in great comfort across perilous seas. The potentialities of beauty in a thing are in direct proportion to the importance or the wonder or the worthiness of the end it serves. As we admire the function, we derive pleasure from its revelation. This is not to apply an ethical standard to art: it is merely to state a simple rule of consequences. A temple to a god or the capitol of a state has greater possibilities of stirring our responsive emotions than a service station or a garage; an airplane can be more beautiful than a trolley car.

But we are also influenced by the degree of success with which a function is performed. Complete adaptation to a humble station in life, perfect fulfilment of a modest destiny, may be far more admirable than meretricious performance in a more exalted role. Hence there are plenty of good service stations and garages which are more beautiful than any number of banal temples and capitols with which our land is dotted, and there are super trolley cars now building which are more beautiful than the older and cruder planes. We respond to the importance of the purpose for which an object is right, but we respond even more definitely to its degree of rightness.

It is easy to overlook the value of this simple homespun self-respect in design. It is done every day: an object's social rating appears lowly, and so its design is falsified in an effort to lift it above its proper station . . .

. . . strange to say a taint of radicalism has attached to the reviving cult of candour and honesty, which maintains that design is evoked from within its object, that it derives its validity from its revelation of function, materials and processes, that while these may not be exalted the design will be still less so unless they are frankly acknowledged . . .

We can appreciate the way that tradition hampers us in spite of our best intentions, if, to take familiar examples, we contrast the rates of advance in the design of airplanes and of automobiles. Airplanes, luckily for them, had no background in history, and so have been able to progress unhindered by deeply grooved habits of thought. To stay in the air at all, the form of a plane must be adapted with great exactness to its function of flying, and so the designers of planes have been forced, willynilly, to advance toward ultimate, unadulterated functionalism. As a result we have produced these thrilling organisms that have more power to move us aesthetically than almost anything else this age has created.

Automobiles, on the other hand, inherited a carriage tradition thousands of years old, and almost any contraption can be made to run on four wheels. So the carriage tradition dominated automobile design and held it back for a whole generation. It was only when the memory of graceful, horse-drawn vehicles had grown dim that we could analyse the automobile's function in its own terms and evolve a rational form to suit it.

Not only tradition, but a number of other influences – fashion, supposed market preferences, mere imitativeness – have all distracted the automotive engineer from pure engineering into irrelevant bypaths and so delayed a solution of his problem. Most effective of all, however, was an impediment for which he could not be blamed, one which exists in practically all fields: the difficulty of accurately defining a function, and definition's habit of retreating before our approach.

When internal combustion engines were first substituted for horses as a means of propelling vehicles, the full implications of the changes were far from revealing themselves. The engine was looked upon as a substitution merely, and it was only as the engine began greatly to surpass the horse in speed

Texaco service station corporate identity, designed by Walter Dorwin Teague (Walter Dorwin Teague Associates)

and endurance that it became necessary to revise the vehicle too in the interest of safety and comfort. A carriage that had been the height of luxury at 10 or even 12 miles an hour became decidedly uncomfortable and unsafe at 20 or 25. Continually greater speeds demanded progressive adaptations, and these speeds, with the increase in driving, demanded a new system of highways. The new highways made still greater speeds practicable, and a network of service stations removed the last restriction from practically unlimited travel. By 1930 we were driving millions of automobiles at cruising speeds ranging from 40 to 70 and 80 miles an hour, with the limit not in sight. Yet these automobiles were still adapted carriages and the engineers had not yet begun to approach a simple and accurate definition of their problem. Gradually we have come to regard the automobile simply as a vehicle to carry passengers over the highways in the greatest possible comfort at speeds limited only by the factor of safety, and at a minimum of expense. Since 1930 an enormous amount of groundwork has been done preparatory to designing a motor car which shall exactly fit this definition, and in this study almost every part of the car has undergone drastic revision: the construction of the frame, the relation of motor to passengers and to driver, the contour of the body as it affects the passing flow of air currents, its construction and its trim, windows, seats, method of springing the wheels and of steering. New factors are continually making their appearance, but in the last ten years we have made substantial advance in clarifying our conception of the automobile's function and in devising a form to fit it . . .

Functional design has advanced most rapidly in fields where the units to be dealt with are smaller and more easily and rapidly replaced. For this reason, air and highway transportation have forged ahead while the railroads have persisted in an almost fossilised state. And while our cities and the housing of our people, the whole scheme of living, working and bartering in fact, compose a vast mass of inertia which can be nibbled at only here and there, the mechanical equipment of daily life in both home and industry has advanced with an astonishing velocity . . .

Function is seldom uncomplicated, and there is always a tendency to seize upon certain outstanding characteristics and neglect others that are less obvious but still essential. It was in this way that automotive engineers so long ignored the need for economy through adaptation of the form to wind resistance while they were engrossed in increasing the speed and power of their cars. Almost all mechanical devices have subsidiary requirements aside from their principal functions. An amateur motion-picture projector must first of all project brilliant and steady pictures, but it should also operate silently and it should never burn the fingers of the operator. A gas range should cook well, but it should also be easy to keep clean, it should bend the cook's back as little as possible, and require a minimum of her attention; and it should fit into the bright, immaculate scheme of the modern kitchen. These secondary requirements have revolutionised range design in recent years, and not the demands of cooking alone, which after all was done quite successfully on the old-fashioned cook-stove.

Certain functional requirements recur again and again in modern mechanical products: ease of operation, for instance, and the appearance of ease of operation. Many products that are really quite simple were left by their producers with such a bewildering appearance of complexity that they terrified the novice and so became actually difficult to operate: today we reorganise these devices to fit their functional definition, suppressing unexplained moving parts and exposed mechanism so that only the essential controls are presented to the operator's eye and hand.

Another recurring requirement is that a product should provide as little lodgment as possible for dust and dirt. This alone has helped eliminate vast quantities of 'ornamental' detail from all sorts of things from typewriters to skyscrapers. Silence is another: if all our mechanical devices were designed without regard to the noise they make, our homes and offices would be unendurable bedlams. Compactness, lightness of weight, durability, automatic operation, low operating costs as well as low production costs, are almost universal functional requirements. To neglect any of these factors is to fail in achieving that ultimate rightness which is the aim of design.

The profession of industrial design in its brief career has developed a technique for analysing the function of a product, and has set up standards for judging functional fitness. It has applied these methods and standards to innumerable objects ranging from domestic and office appliances to automobiles, ships and trains; and to industrial and commercial buildings, offices and shops, and the expositions in which industry explains itself to the public. It has set up a procedure which must be followed in devising housing that shall approximate our cars in efficiency and low cost; in developing a system of transportation whereby our planes, trucks, cars, highways and railroads can be co-ordinated into one smoothly functioning system for the distribution of goods and people; and in creating forms of . . . living which shall supersede the present unworkable town and city plan.

None of these problems will be solved automatically. They cannot be left to mere natural evolution. It is true that in certain products, especially those we have cited that move at high speed, right form evolves by necessity: any deviation from purely functional form is disastrous and the engineer is forced into functional design because he has no choice. In machine parts, too, the same compulsion operates and we have the marvellous abstract perfection of these forms – a beauty which makes Mr Brancusi's sculpture seem somewhat trivial. But while the parts may be beyond criticism, the assembled machine may not be: as the form grows complex the possibilities of variation increase and often arrive at confusion.

Perfection of form is achieved automatically only when necessity leaves no choice in the matter. In all other cases conscious design must supplement engineering, and these other cases are the vast majority of all we do. The perfection of these naturally evolved functional forms proves that exact functional adaptation is the only proper aim of all design. In no other way can such vital distinction, such moving significance, such satisfying rightness be evoked. And exact functional adaptation of course implies adaptation to materials and processes of manufacture as well as the unification of the whole. It is impossible to disentangle these phases of the design problem.

From 'Fitness to Function', Chapter 4, Design this Day – the Technique of Order in the Machine Age *Harcourt, Brace & Co, New York, 1940, pp 50-66. Reprinted with the permission of Walter Dorwin Teague Associates.*

Steinway grand piano, designed by Walter Dorwin Teague, 1939 (Walter Dorwin Teague Associates)

Harold Van Doren 1940

Buttons for Levers, the Borrowed Streamline

Of all the leading American industrial designers who published books about their craft, it is Harold Van Doren who paid most attention to the engineering reality of the new vocabulary of shapes which were to become the iconography of consumerised American industrial design.

That Van Doren features in this discussion a hypothetical laundry equipment company illustrates the fact that, by 1940, streamlining had become an industrial designer's metaphor. Technically, streamlining is the science of studying how bodies pass through fluids. Most usually it is applied to bodies passing through air and it was the increasing popularity of aeroplanes and air travel during the 1930s that initially introduced the public to the smooth forms which function made necessary. Soon it was realised that the streamlined forms which nature forced on birds and fish and necessity required for aeroplanes and cars were intrinsically attractive and they were used on objects and machines, like toasters, vacuum cleaners and refrigerators, which were never intended to pass through any fluid at high velocity.

This metaphorical use of streamlining – itself most often symbolised by three parallel chrome strips, on trains, typewriters and automobiles – became the trademark of the first generation of American industrial designers. It also required industrial designers to blend all the disparate elements of a complex, manufactured object into a uniform whole. It was both the stylistic innovation which caught the public's imagination, and the deception which infuriated serious critics of industrial art.

Van Doren's discussion is, in this context, a most interesting one. His engineering training equipped him fully to understand the manufacturing processes involved in making streamlined shapes, but although this same technical awareness made Van Doren contemptuous of sham, in his book, as in this chapter, commercial considerations are uppermost.

The scene is the office of the Vice-President and Sales Manager of the Peerless Laundry Equipment Corporation. The Vice-President is speaking.

'Gentlemen, this is Mr Blank. He has been engaged by our company to streamline our line of laundry tubs.'

Mr Blank is one of that mysterious yet over-publicised fraternity of industrial designers. He is being introduced to other members of the company: the vice-president in charge of production, the chief engineer, the plant superintendent.

Streamlined laundry tubs! What is poor Blank to say? He knows that the substitution of a few radii or soft curves for angles and sharp edges bears about as much relation to genuine streamlining as knitting sweaters does to the science of hydraulics. But the client calls it streamlining.

Streamlining has taken the modern world by storm. We live in a maelstrom of streamlined trains, refrigerators and furnaces; streamlined bathing beauties, soda crackers, and facial massages.

True streamlining is not nearly so important, nor so frequently met with in problems of industrial design, as the general public probably thinks. But it is a phenomenon no designer can ignore and no modern book on design can afford not to discuss.

The manufacturer who wants his laundry tubs, his typewriters, or his furnaces streamlined is in reality asking you to modernise them, to find the means for substituting curvilinear forms for rectilinear forms. He wants you to make cast iron and die-cast zinc and plastics and sheet metal conform to the current taste, or fad if you will, for cylinders and spheres or the soft flowing curves of the modern automobile in place of the harsh angles and ungainly shapes of a decade ago.

He expects, too, that unnecessary exposure of mechanical parts will be eliminated, that buttons will be substituted for levers, and control panels and dials will be organised into simple and easily read groups, tied in wherever possible with related elements of the principal forms.

In 99 cases out of 100 this procedure is utterly unrelated to genuine streamlining. Then why does the client refer to it as such? And why should designers be streamlining everything, from lipsticks to locomotives?

Simply because, in the unbelievably rapid growth of the American language, words soon lose their specific and restricted meanings and assume a general significance embracing far wider fields than originally intended. The streamlined laundry tub today means the modern laundry tub. It means the very latest up-to-the-minute laundry tub that can be bought, something so ultra-advanced that it is almost more of the future than the present.

In our rapidly shifting language the life of such words, at least in their popular form, is often brief. The vogue for 'streamlining' dates from about 1934. And yet the Oxford Dictionary, that ultimate authority on the history of English words, tells us that it first appeared in print as a term in hydrodynamics in 1873. In 1906 some anonymous designer described it as follows: 'A "line of motion" or "streamline" is defined to be a line drawn from point to point so that its direction is everywhere that of the fluid'. The word used in the attributive form we know today, 'streamlined', saw the light of day for the first time in 1909. Motor manufacturers trying to characterise the increasingly sweeping lines of their cars used the word in this sense soon after. I distinctly remember as a boy having a small friend boast to me that his father had bought a 'streamlined' Hudson. It was a touring car, and for the first time a manufacturer had built an automobile in which the lines of the sides swept from hood to rear without the interruption of projecting seat backs.

But the word didn't take. More than two decades were to pass before 'streamliner' became synonymous with the modern railroad train and before manufacturers were to summon their salesmen to a 'streamlined convention'. Nobody can tell how long its current vogue will last, or when it will subside again to its original and more technical level of meaning (again it is the Oxford Dictionary speaking): 'that shape of a solid body which is calculated to meet with the smallest amount of resistance in passing through the atmosphere'.

Figure 1

Nonfunctional Streamlining

Since streamlining in its minor or nonfunctional sense – the substitution of radii and fillets for sharp angles and corners – has become such a factor in modern design, it will be necessary to discuss ways and means of accomplishing these effects. But first you must understand some of the manufacturing processes involved . . .

Fabrication limits you to purely geometrical forms. Stamping, on the other hand, permits much greater latitude in the development of what, for want of a better name, we shall call 'freehand forms', that is, forms curved in two or more planes at the same time. Fabrication methods cannot produce true streamlined shapes; stamping techniques can. Spinning is capable of forming sheet metal with a freehand curve in one plane and true circles in a plane at 90 degrees to it.

Various methods of casting and forming metals, plastics, and rubber (sand casting, permanent mould casting, plastic moulding, rubber moulding) – in fact any of the methods whereby the

Figure 2   *Figure 3*

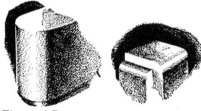

Figures 4-7

Figure 8

material takes its form after being introduced in a molten form, or a form which becomes molten or plastic under heat or pressure or both – are also capable of producing true streamlined shapes.

Sketch 1 has been fabricated. The main part of the form, which might be half of the casing for a water heater, was shaped in rolls. The bead near the top was put in on a small rolling machine equipped with cylindrical dies. The 'revealed' edge at the bottom was made in the same manner, being offset just the thickness of the metal so that a base could be fastened to it and produce a flush joint.

It must be pointed out that this form could have been made on a stamping press. If it were stamped, the piece price would probably be less, because hand labour would be largely eliminated, but the die cost would be large. The process used would thus depend entirely on the number of units manufactured. Note carefully that the main body is a half cylinder, *curved in one plane only*, which distinguishes it from a typical stamped form.

Now examine sketch 2. Let us suppose that it is the base for an electric fan or perhaps a domestic food mixer. It is an imitation of a functional streamlined shape, borrowed from aeronautical practice. This form could not possibly be made by fabrication methods. If made of sheet metal it would have to be stamped. It could also be manufactured by any of the casting or moulding processes, but spinning, here, would be impossible.

The next sketch, 3, shows another genuine streamlined form, the tapered rear end of a headlamp, or the shell of a hair-drying machine for instance. It could be made by spinning, stamping, casting, or moulding, but would be impossible to fabricate, in the sense used above, because again it is curved in more than one plane.

Fabricating sheet metal, for all its limitations,

offers possibilities that have not been fully exploited. Let us examine a few of the basic shapes that can be made by fabrication methods, and then have a look at a few details.

Figures 4-7 indicate a few of the directions that might be taken. They are not intended to represent any particular product, being presented as abstract forms only. With a little ingenuity and the proper equipment, any one of them can be fabricated. Certain subtleties, such as the 'peaked' treatment already mentioned, might also be introduced, but it would be well to use this treatment sparingly and only on plane surfaces. You cannot push this peak around a curve without changing the technique to stamping.

Corner Treatments

Now let us look at a few simple corner treatments. Certain types of heating equipment, for example, may call for light sheet-metal casings, squarish in shape. Cost is paramount; volume therefore does not warrant the use of expensive stamping dies. Yet the client wants some evidences of streamlining. Perhaps the treatment permitted is merely on the corners and edges; a little striping, a few lines of chrome beading, colour selection, and a nameplate are all that cost will permit.

What shall we do? There are good and bad ways of rounding edges and of treating corners where three planes meet. In Figure 8 we see the corner of a small domestic air conditioner treated as the average sheet-metal shop would make it if streamlining had not come into the picture. There is nothing wrong aesthetically with a corner like this, *as a form*, but in fabricated metal it has a tendency to make the entire casing look tinny and cheap. Let us see what we can evolve out of this uncompromising geometry.

Our first thought is to put a radius on all three edges, where the planes intersect. This produces a 'ball corner', as seen in Figure 9. There are several ways of making it, such as notching out the corners and inserting a piece made on small corner dies, then spot-welding it into position; or gas-welding in a corner piece and grinding off the excess metal. As design it is quite all right, but not particularly interesting. The trouble is that all of the radii are the same; there is no variety of form.

We can enlarge the radius of the vertical edge only as in Figure 10. This will help to give character and variety. It creates, however, a somewhat peculiar shape for the corner piece. If the entire top could be a shallow stamping, everything would be well, for, by offsetting the metal on the side-and-front piece with a small roll, we could obtain a flush joint.

Enlarging the radius on the horizontal edges is another alternative as in Figure 11, but it gets us into trouble. The juncture of a small radius on the vertical edge with two larger ones on the horizontal plane produces a kind of peak and results in a freehand, three-dimensional form not particularly pleasant to the eye and difficult to make.

Perhaps we should throw caution to the winds and try something more radical. The limitations of fabrication are such that we cannot dome the top and indulge in any freehand forms. But we can do two things: we can greatly increase the radius of the front edge, producing a soft roll from front to back; or, we can enlarge the radius at each side, again producing a markedly different effect. Refer to Figures 12 and 13. Both of these would require the use of a radius former, unless stamping were resorted to. Which scheme will be chosen depends on the other minor forms involved, the mechanisms inside, and the location of the product in its final installation. The treatment chosen from any of the above suggestions will depend also upon the differences in cost, determined when the cost accountants have had their innings and each operation has been broken down and analysed to the last penny.

The chief purpose of this demonstration is to indicate the *form* possibilities inherent in the simple intersection of three flat planes, manufactured by one of the most limiting of all production methods. Much more intricate designs could be developed by combining fabrication with stamping techniques, or by using die-cast or sand-cast parts to obtain the desired corner forms.

True streamlined forms can never be decided on paper. Here modelling to scale or full size is an absolute essential to success. Work from the start in three dimensions. I do not mean to say that sketching with the pencil should be abandoned entirely, but it can never give you a true picture of freehand forms.

The real field of true streamlining is transportation, especially the airplane, although it has hydraulic applications as well. In both of these fields it is functional. It has been borrowed in many forms by industry for application to static objects, where its employment is due largely to popular fancy. This is not the place to debate the merits of borrowed streamlining, although if it helps sell merchandise that should go a long way toward justifying its use.

From 'Streamlining', in Industrial Design – a Practical Guide *McGraw-Hill Book Company, London and New York, 1940, pp 137–148. Copyright 1940 by McGraw-Hill Inc. Used with permission of McGraw-Hill Book Company.*

Figure 9

Figure 10   *Figure 11*

Figure 12   *Figure 13*

Raymond Loewy 1945

Delivering the Goods

This is a full transcript of a letter published in The
Times *during November 1945, at a time when British
industry was considering its own renewal and regeneration after the war. It is especially interesting as a
documentary piece: Loewy's comparison of British industry with its American counterpart would be as relevant
today as it was in 1945. The relative lack of expertise,
standardisation and series production in British industry
are still serious faults which bedevil progress.*

*Raymond Loewy, as the most popularly successful of
all American industrial designers, has maintained a
sophisticated dualism throughout his working life. He
believes in the importance of beauty, but it is beauty as
both a commercial and an aesthetic phenomenon. That
his design philosophy – succinctly suggested by the title of
his autobiography,* Never Leave Well Enough Alone
(1951) *– was the right one is indicated by the fact that in
1945 Raymond Loewy Associates was on retainer to no
fewer than 79 American, British and Swedish corporations as design consultants. At the beginning of
industrial curtailment on American entry into the Second
World War, products for whose design Loewy was
responsible grossed $750,000,000. By 1946, with pro-
duction resumed, that figure had risen to $900,000,000.*

Raymond Loewy had been put in touch with The
Times *with the idea of producing an article on an American designer's view of British industry. His submission
was found to be too short to use as a feature and was
eventually published as this letter.*

Industrial Design — The Aesthetics of Salesmanship — An American View

It is a generally accepted fact that the economic
future of the UK is closely related to its export
trade. Unless it is increased by at least 50 per cent,
the British standard of living will be seriously
threatened. The urgency of the matter requires that
this flood of consumer goods should bring with it
every advantage of quality, low cost, and the visual
appeal of modern design.

Avoiding the exquisite intricacies of the problem, may I express an American industrial
designer's views on the subject? In so doing I might
be of some assistance to my British manufacturing
friends; for I am convinced that it is essential that
we Americans should continue to co-operate with
England as closely as we did during the war. Our
interests, in the long run, are parallel.

In America the industrial designer has developed
a design technique that has been successfully tested
in the field during the past 15 years, and American
products most generally carry with them the
advantage of collaboration between the professional consulting designer and the industrialist.
This partnership with industry and our philosophy
of design are based on securing greater sales appeal
– increased trade. I have heard much here in England about the aesthetics of design. This to me is
strange. We seem to find that the aesthetics of an
industrial product will take care of themselves
automatically after we have provided a balance
between function, simplicity and utility. In the
average manufacturer's mind the word aesthetics
has annotations of philanthrophy and culture after
office hours. Industrial design did not become
standard practice in America because of aesthetics.
There were far too many other good reasons for

using it – increased sales appeal of products and increased trade. And the final result is no less beautiful for having been designed on sound, unemotional business lines.

The industries which have benefited most from design are just those which formerly were in greatest artistic disrepute. Here is where the industrial designer can be of immense help to the export-minded manufacturer. The professional designer schooled in manufacturing methods can and does aid in achieving maximum simplicity both of manufacturer maintenance, and operation, with a resulting increase in the demand for these products. The high quality of British products is known throughout the world. There are, however, many occasions when it is difficult to detect quality or the lack of it through the very nature of the product, the ersatz looks as good as the original (until you learn later). The high quality of many items is often obscured by such trivia as an irritating colour scheme of a disorganised appearance. Problems such as this are a professional matter with the industrial designer. Quality can and should be made apparent, for British quality will not remain unchallenged. In America, for instance, hundreds of thousands of unskilled workmen have been trained during the war in precision-manufacturing. This will tend to improve . . . US products.

American industry is tending to standardise and reduce the number of models manufactured in any given line. This affords higher unit-production and consequent lower cost. It brings the products within reach of an enormous section of the population in the lower income strata. Never before were these people able to afford all sorts of household labour-saving devices that would improve their standard of living. 'De luxe' models will continue to be built for those who can afford them and who are willing to pay the cost penalty inherent in low-scale production.

Industrial design, as I know it, 'delivers the goods'. It is a serious profession which combines good taste, technical knowledge, and common sense. In the case of my own organisation, it is taken seriously by over 75 corporations who are planning to build during 1946 over £200,000,000 worth of products designed with our assistance. Their conception of aesthetics consists of a beautiful sales curve shooting upwards. This realistic approach is certainly successful. May I suggest it for the consideration of the industrial designers of Britain, who must share in the tremendous responsibility of . . . industry in today's great problem?

From 'Industrial Design – The Aesthetics of Salesmanship – An American View', letter to The Times *(London), 19 November 1945. Reprinted with the permission of the author.*

Hupmobile sedan, designed by Raymond Loewy, 1934 (Raymond Loewy International Inc)

Lucky Strike 'White' cigarette packaging, designed by Raymond Loewy, 1942 (Design Council)

Hallicrafters radio receiver model SX42, designed by Raymond Loewy, 1946 (Design Council)

Good Design is not a Luxury

This was the very first article published by Design *magazine, the official journal of the Council of Industrial Design (later the Design Council) whose policy Gordon Russell did so much to create.*

Gordon Russell, craftsman and administrator, was a furniture designer and Managing Director of Gordon Russell Ltd of Broadway from 1926 to 1940. During the Second World War he played a major part in the creation of utility furniture. He became Director of the Council of Industrial Design in 1944, and was an original committee member involved in the preparation of plans for the Festival of Britain.

His article, reprinted here, is a classic statement of Council of Industrial Design policy, a careful blend of practical aesthetics and puritanism, claiming sources both in the craft tradition and in the work of progressive Continental design theorists. Arguing the established case of maintaining truth to materials, and condemning the application of inappropriate forms to inappropriate objects – for instance, streamlined refrigerators – Russell has succinctly described post-war British design philosophy. At the same time as it prompted the austere ideals of the 1930s, it also condemned – with some severity – the panache of American stylists whose work also took form in the pre-war period.

Many people speak of good quality as if it were made up of good workmanship and good materials alone: but without good design it is impossible to make the most of these qualities. *Good design, indeed, is an essential part of a standard of quality.* Without it, the manufacturer cannot give the best service, through his products, to the consumer – to the community of which he is himself a part and from which he derives his livelihood.

What does the consumer demand in a manufactured article? He demands something which is well made of good and suitable materials, which does its job efficiently and gives him pleasure, at a price he can afford to pay. So the first design question is 'Does it work?' You have all seen clocks with hour and minute hands so similar that it is not easy to tell the time, teapots which do not pour well, kettles which burn your hand, handles which pinch your fingers. These are all examples of bad design, and there are many others.

Though 'Does it work?' is a good approach to design, it will not take us all the way. Even where science can virtually define shapes, as in the case of the aeroplane, one of our most famous aircraft designers has said: 'I like a thing to *look* right. If it doesn't, although I may not be able to prove scientifically that it is wrong, I have often found out later that it is.' Here is a practical application of aesthetics which may seem strange, yet I can think of many others: dark and dirty factories, ugly dull-coloured machinery, unpleasant lettering, inefficient packaging, disregard of shape and texture and colour – all forms of bad design – will be taken more seriously in the future, because they are deterrents of production and sales.

Good design always takes into account the technique of

production, the material to be used, and the purpose for which the object is wanted. You cannot get satisfactory results by designing for hand production and then turning over the same design to the machine. Nor can you design for one material and then make the object in another. The wax candle was the best form of illuminant in its day, but as a prototype for electric light it leaves much to be desired. The development of the railway-carriage was arrested for several generations because it was thought of as a series of stage-coaches: not until the new idea of inter-communication was grasped did the corridor make possible improvements in design (larger windows, less draughty compartments, restaurant cars, lavatories).

The materials to be used for any product should be chosen with care, not only to be economical from a manufacturing point of view, but to wear well in use. In plastics, for instance, it is no good making a tea-strainer of cellulose acetate, which will not stand up to hot water.

In places where easy cleaning is essential, a smooth surface must be used, but a thoughtful use of rougher textures can often give variety and interest elsewhere. Here nature is a great teacher.

Then we come to the question of ornament. Today, much of the beauty we associate with the machine springs from intense preoccupation with the best way of achieving a given result by sparing use of suitable materials rather than by added decoration. But from earliest times, men have loved to decorate the things they made with simple geometrical patterns, pictures of animals, trees and so on; and the evolution of a contemporary decorative style is a problem we have yet to solve. Many articles in plastics try to give an impression of having been carved by hand, whereas they are moulded in a press; the so-called carving is lifeless.

Refrigerators, which remain stationary, are streamlined as if they were aeroplanes or ships. And how many objects have three zigs up and three zags down plastered on them? These design clichés are not the right answer to a human need. We in Britain cannot afford to be left behind in this aspect of industrial design. It is bad business if our customers think of us as being uninterested in the *look* of our goods, but I could give you many instances where they have that impression today.

To any design problem there are many possible solutions; there is no one perfect solution, and sometimes, as in the design of a flower vase, there are hundreds or even thousands of shapes which would do the job. The designer is a person who, among other things, is always studying shapes and so is able to evolve or select one which not only works well but means something. This applies equally to form and colour: the designer is able to give shape to aspirations which all of us possess, but which we have not the training to create for ourselves.

We can learn something of the meaning of good design by considering what it is NOT. To clear away one elementary misconception, let me say that *good design is not precious, arty or highfalutin'.*

Again, it is not a luxury that enters into the more expensive end of a trade only. It is true that new styles not infrequently start in luxury markets; indeed, I believe that luxury trades perform an indispensable function by enabling experiments of all sorts to be tried out in a small way. But mass production so spreads the cost that there is no reason why well designed things should not be available for everyone to buy. The idea that only wealthy people like well designed things is as false as that they are the only people to get pleasure from looking at flowers, listening to music, or reading

"Does it work? Does it look right?" There are still far too many products for which the answer must be No

Artwork from Design *magazine, vol 1, no 1, 1949 (Design Council)*

Densitometer manufactured by Ilford Ltd, c1949 (Design Council)

Airspeed Ambassador airliner; thought, in 1949, to be the most streamlined of all aircraft shapes (British Aerospace)

Shaw. Equally false is the notion that because a thing is low in price it cannot be of good quality.

Good design is not something that can be added to a product at a late stage in its planning or manufacture. It is fundamental. Before starting on a job, any designer worth his salt makes a complete survey of the problem. The manufacturer who is not prepared to place all the relevant information at his disposal cannot expect to get the best results. A clear statement of the problem is essential to its satisfactory solution. In what market it is proposed to sell the product? At what price? Against what competition? How will it be marketed? How packaged? What materials are to be used? What machines? A detailed survey at the outset will save much trouble later.

There used to be many people who thought that an architect was employed to ensure that the elevation of a building should be in a given style, but in fact a good elevation grows out of a good plan: the architect's true function is to grasp the needs of a client – needs which he may not be able to state precisely – and crystallise them into a workable plan that is economic to build and pleasant to live, work or play in. The industrial designer is, as it were, another kind of architect – the co-ordinator in a team of specialists. He must, by the nature of his job, work as one of a group of technicians. At every stage of the work he must be closely in touch with other specialists, saying to one: 'Is this the best way to machine this job, or shall we cut the corner?', to another: 'What material shall we use here?', to a third: 'Is this likely to give trouble in the packing department?', to a fourth: 'Could you sell this for £16 10s 0d?' and so on. A designer calls on the experience of a great number of people in the firm for which he is working – works manager, sales staff, foremen, advertising and costing men,

research staff and so on.

Research into design is a part of industrial research which in the past has been sadly neglected. Like research as a whole, it can only be tackled by adopting a policy which goes steadily ahead over a period of years; you cannot expect each year's results to pay for themselves. There are no short cuts. A firm cannot pack up bad design on Friday night and start churning out good design on Monday morning. It is not so easy as that; it is necessary to change the point of view of a number of people in the organisation. However, it is *not* necessary to start in a big way and perhaps throw up assured profits in the hope of securing others which may not mature. The design department that starts in a small way today is likely to grow until it becomes the mainstay of tomorrow. Good design does not sell itself but it can be made a strong selling point. Its prestige value is great – and growing: the standard of public taste is rising.

An approach to design through horse-sense enables any intelligent person to appreciate what the designer's problem is – and that is what we need. We don't expect everyone to become expert designers; that is neither possible nor desirable. We cannot all become accountants, but we can learn enough to read a balance sheet. We cannot all become conductors but we can learn to appreciate music – and remember, no conductor could give his best to an audience of deaf mutes; there must be collaboration. It is the same with design: a public which possesses critical standards is essential if design is to be as good as it might be.

Sometimes we hear it said that there is no such thing as good or bad design, that there are no real standards by which design can be assessed, that it is just a matter of personal taste; or that because an article sells in great quantities it *must* be well designed. Sometimes, too, we are told that the subject is not a very important one; that hard-headed business men cannot be expected to waste their time on what they think is purely a question of aesthetics, and so on.

I have heard such criticisms on many occasions in the past; they are becoming rarer today. More and more people are realising that the question of industrial design is important to industry, and indeed to every citizen. In 1944 the Coalition Government, with all the preoccupations of the closing months of war pressing on it, found time to set up the Council of Industrial Design, largely because it was felt that design was vital in our post-war export trade.

In 1951 the Festival of Britain will provide an opportunity to show that this country, which once led the world in design, is ready to assume leadership again. It is the Council's task to select a variety of products to be shown in the Festival: it is industry's responsibility to ensure that a first-rate range of goods is available. A permanent grading-up of standards, as distinct from a short-lived attempt to produce 'stunt' designs, can do much to ensure the future prosperity of our country. 'British made' ought always to mean well made of sound materials to a good design.

From 'What is Good Design?', in Design (London), *vol 1, no 1, January 1949, pp 2-6. Reprinted with the permission of* Design *magazine.*

Sharp Practice

Reyner Banham, who was born in 1920, has been, at one time or another, an apprentice with the Bristol Aircraft Company, a journalist and an academic. A one-time student of Nikolaus Pevsner, today Banham is the most respected international authority on architecture and design.

As a staff writer on the Architectural Review *during the 1950s, Banham created the second generation of the history of modern architecture in a series of stylish articles which culminated in the publication of his book,* Theory and Design in the First Machine Age *(1960). This extract from his article of April 1955, called 'Machine Aesthetic', is one of them.*

A concern with images and imagery dominates Banham's writing about art. This is no coincidence; Banham's own interest reflects the contemporary preoccupations of the Independent Group, the British Pop Artists who were gathered around the ICA. Banham was one of its leading members.

This article was an attempt to define the modern architect's and artist's infatuation with machines, viewed from an historical standpoint. There is a rich and interesting paradox inherent in this subject. At the same time that architects and painters were turning to the machine as an exemplar of good, rational design which would inspire them, industrial designers and engineers were anxiously trying to humanise the machine by making it beautiful. Le Corbusier, for instance, spoke favourably of the rational necessity which made Ettore Bugatti's famous car engines so beautiful. We know now, however, that so far from being 'inevitable', Bugatti used to make wooden mock-ups of his motors to test them for their visual quality before going into production!

Architects are frightened of machinery, and have been so ever since engineering broke loose from the back pages of Vitruvius and set up on its own. Even where they have paid lip service to 'the Machine' they have paid it to a simulacrum of their own invention; they have been prepared to do business with it only on their own terms, and only with those aspects of it which lie closest to architectural practice. When Adolph Loos hailed engineers, in 1898, as 'our Hellenes' he may have been speaking far less metaphorically than we have so far suspected, since his fellow-Viennese Otto Wagner had already looked to engineers to restore 'powerful horizontal lines, such as were prevalent in Antiquity'. Similarly, to admire the ranked cylinders of a grain silo was not so great a feat of visual athletics to those who had been trained to accept the lumpish columniation of the temples at Paestum.

The 'Machine Aesthetic' of the pioneer masters of the Modern Movement was thus selective and classicising, one limb of their reaction against the excesses of Art Nouveau, and it came nowhere near an acceptance of machines on their own terms or for their own sakes. That kind of acceptance had to wait upon the poets, and particularly Marinetti, whose Futurist Manifesto of 1909 not only opens with the first car crash in European literature, but contains the pioneering value judgment: '. . . a roaring racing car, rattling like a machine gun, is more beautiful than the Winged Victory of Samothrace'. Such an opinion could only be expressed in the rare atmosphere of pure poetry, and the grey eminences of architecture continued to square up to the problem slowly, picking their ground with care. They had committed themselves to a machine aesthetic of some sort, and the words of Marinetti must have stuck like barbs in the flesh, but to accept his viewpoint would have been to let go of architecture as they understood it. The selective and classicising approach had therefore to continue – as one sees in the pre-1914 factories of Peter Behrens, which are a long step toward a mechanistic architecture, but remain, for all that, neo-classic temples in form and silhouette.

Both at the Bauhaus and in the circle of *L'Esprit Nouveau*, this approach continued, in however disguised and complicated a form, making it possible to bracket together architecture and machinery with the least mental strain for the architectural side. Bauhaus masters may expostulate that they set up the design of machines as an exemplar of method, and did not posit any formal resemblance between machinery and the bare, spare rectangular architecture they produced, yet Malevitsch, in *Bauhausbuch No 11*, hopefully says of his own filleted and rectilinear aesthetic: 'thus one may also call Suprematism an aeronautical art'.

But Bauhaus theory, as we receive it now, is very fragmentary, and one should be chary of invading this difficult field until the group psychologists have been into it, for it is the Bauhaus *atmosphere* which needs to be studied. *L'Esprit Nouveau* is a more immediately rewarding field, with its neat corpus of signed articles and all pseudonyms known, and its effect upon the growing concept of the Machine Aesthetic is clearer and easier to follow. In this body of articles one will find, *in extenso*, the manipulation of the superficial aspects of engineering in the interests of a particular conception of architecture, and in those articles which were later published as *Towards A New Architecture* one can see Le Corbusier advancing a view of machinery which progresses shortly from special pleading to false witness . . .

His background in his native Chaux-de-Fonds was watchmaking, still in an eighteenth-century condition compared with the production-line industries; his architectural apprenticeship was with Behrens and Auguste Perret, both old-time classicists and the latter a self-confessed reactionary whose model in concrete was wooden framing; and his 'industrial experiences', if indeed they were with the Voisin company, were in an aircraft industry which was barely out of the box-kite phase. These facts remembered, his naive belief that machines are by their very nature highly finished can be understood – a watchmaker could hardly think otherwise – and so can the extraordinary penetration with which he views the pre-history of flight: 'To wish to fly like a bird is to state the problem badly . . . to seek a means of support and a means of propulsion is to pose it properly.'

But can such naivety explain the crooked argument of the chapter in *Towards a New Architecture* entitled 'Automobiles'? Its crookedness is disguised by the fact that the argument is partly verbal and partly visual. The hinge of the verbal argument is the virtue of standardisation; the hinge of the visual is the confrontation, sustained over several pages, between automobiles and the Parthenon, and the totality has been read by two generations of architects and theorists as meaning that a standardised product like a motor car can be as beautiful as a Greek temple. In its context that is how it must be read, but the *tertium comparationis* of the argument is a disingenuous pretence – none of the motor cars illustrated is a standardised mass-produced model; all are expensive, specialised, handicraft one-offs which can justly be compared to the Parthenon because, like it, they are unique works of hand-made art. Mass-produced vehicles . . . are not allowed to sully these classicist pages.

Above: Temple at Paestum, 600-550 BC and Humber motor car, 1907; below: the Parthenon, 447-434 BC and Delage Grand Sport, 1921. From Vers une Architecture *by Le Corbusier, 1923 (© by SPADEM Paris, 1978)*

Cover of Die Gegenstandslose Welt, *Bauhausbuch no 11, designed by Kasimir Malevich, 1928 (*The Bauhaus *by Hans Maria Wingler)*

Engine of a Bugatti type 35 motor car, designed by Ettore Bugatti (Hugh Conway)

Naivety? Sharp practice? Or wishful thinking? A certain aesthetic *parti-pris* is undoubtedly there; a desire that certain wishes should come true; that architects should in reality be able to assume the moral stature of engineers on whom, in the opening chapter, Le Corbusier had wished the virtues of the Gothic Craftsman and the Noble Savage: 'Engineers are healthy, virile, active and useful, moral and happy.' It would be difficult to write this kind of nineteenth-century fustian with one's tongue in one's cheek, and an intention to deceive would be very difficult to maintain against him, since he does not even seem to see how the parenthetic caption to one of his motor car illustrations, *Carrosserie Ozenfant*, undermines his argument by conjuring up the presence, not of a moral and virile mass-production engineer, but either a luxury carriage builder, or his artist son who was Le Corbusier's pictorial collaborator. This same Amedée Ozenfant was later to depth-charge the whole argument by pointing out that 'M Ettore Bugatti as well as MM Voisin Farman and the Brothers Michelin had all been art students' so that even Voisin, the patron of *L'Esprit Nouveau*, was not an untainted engineer happy in his morality and usefulness.

Wishful thinking cannot be ruled out; nor a desire to find support for one's aesthetic prejudices on some human activity which can be admired without qualification, and this clearly opens the way for extensive self-deception. The particular wish-confusion which lies at the bottom of this complex structure of deception and distortion is easiest to identify in *La Peinture Moderne*, an otherwise excellent book which Le Corbusier and Ozenfant made out of another series of *Esprit Nouveau* articles which had appeared in parallel with those on architecture. Here they say of Purism, their own

style of painting: 'Le Purisme a mis en évidence la Loi de la Sélection Mécanique', and this law, which they clearly intend to share the status of Darwin's Law of Natural Selection, they explain as follows: 'It establishes that objects tend toward a type which is determined by the evolution of forms between the ideal of maximum utility, and the demands of economical production, which conforms inexorably to the laws of nature. This double play of forces has resulted in the creation of a certain number of objects which one may call standardised' and, the argument runs, are therefore good, and have been selected as the Purist's subject matter. The relation between the law and the paintings only concerns us in one respect here: the authors have set up an abstract model of the design process in mass production, and the paintings will show us what class of objects we have to interpolate as the last term of the proposition in order to test its truth.

These Purist objects prove to be bottles and jugs, pots and pans, glasses and pipes, of forms which approximate to the cylinder, sphere and cone which had been canonised in post-Impressionist painting, regular geometrical forms with simple silhouettes. If we make these the last term in the Ozenfant-Corbusier model of the design process, we get a proposition of this order: Objects of maximum utility and lowest price have simple geometrical shapes. To most architects this proposition would appear watertight, but to most production engineers it would appear too abstract to be useful, and demonstrably false in its outcome.

To them, Utility, in the Rationalist sense which the authors clearly intended, is a marginal factor – only one among a number of other factors bearing upon sales. When the American Ford Company issued a questionnaire to discover what qualities buyers sought in cars, most answerers headed their lists with such utilitarian considerations as road-holding and fuel consumption – a result which sales-analysis did not support – but when asked what *other people* looked for, most headed their lists with chromium plate, colour schemes and so forth. To manufacturers, utility is a complex affair which, in certain products for certain markets, may require the addition of ornament for ostentation or social prestige. Similarly, the demands of economic production do not, as the authors of the model supposed, follow the laws of nature, but those of economics, and in fields where the prime factor in costing is the length of the production-run a simplicity, such as would render a handicraft product cheaper, might render a mass-produced one more expensive if it were less saleable than a more complex form. High finish, too, is another Purist mirage, for the quality which interests engineers is not finish but tolerance – the factor by which a dimension may vary from the designed figure without injurious effects. This renders high finish a purely negative characteristic, and where it is extensively applied to any object it is nearly always the product of handicraft labour, and has some bearing on sales – for reasons of consumer preference, as in luxury cars and watches, or performance, as in airliners and surgical equipment.

All these qualities then – summed up as simplicity of form and smoothness of finish – are conditional attributes of engineering, and to postulate them as necessary consequences of machine production was to give a false picture of the engineer's methods and intentions. But such a picture was clearly of the greatest polemical utility to the Purists in their search for a justification of their aesthetic preferences. It is also clear that they were not alone in this, for the Machine Aesthetic was a world-wide phenomenon, nor was its mythology noxious at the time, for it answered a clear cultural need in offering a common visual law which united the form of the automobile and the building which sheltered it, the form of the house, the forms of its equipment and of the art works which adorned it. Nor – and this is the heart of the matter – was its falsity visible at the time, for automotive, aeronautical and naval design were currently going through a phase when their products did literally resemble those of Functionalist architecture. The Intelligent Observer, turining from one set of smooth simple shapes to the other, would see apparent and visible proof of the architect's claim to share the virtues of the engineer.

But these days were numbered. Already, in 1921, aeronautical design was launched upon a train of development in which a third quality, not mentioned in Le Corbusier's original Support-Propulsion formula, was to dominate the field. That quality, now common to all forms of motion research, was Penetration, and in pursuit of ever better factors of penetration typical aircraft forms were to engorge their structure, and turn from complex arrays of smooth simple shapes, like those of Functionalist architecture, to simple arrays of mathematically complex forms. At the period when the crisis of this development was reached in the early 1930s, with the general change-over to monoplane configurations and retractable landing-gear, the process was doubled in the field of automotive design, where the liberation of bodywork from horse-and-buggy concepts is aptly symbolised by the way in which the Burney Streamliners rendered Walter Gropius's architecturally conceived Adler cars obsolete in a bare 18 months. Within another 18 months the slump had done some rough surgery on the motor industry, lopping off its weaker members and leaving a few giants battling for mass markets and low production overheads. The consequent philosophy of long manufacturing runs and rapid repeat orders led inevitably to vehicles which were very different from the hand-made art works which had graced the pages of *Towards a New Architecture*. It was not merely that pressed steel technology works most efficiently with broad smooth envelope shapes, but also that the need to chase the market led to the rapid evolution of an anti-Purist but eye-catching vocabulary of design – which we now call Borax.

The tone of architectural response to these developments was to complain that machine designers were failing in their task – a tone which had been set by a caption writer in *Cahiers d'Art* as early as 1926, who had accused the engineers responsible for an artless coaling-gantry of the normal splay-legged type of being 'soaked in Romantic Expressionism' – an inexplicable performance unless one believes the Law of Mechanical Selection, and shares its preference for elementary solid geometry. Under the turbulent conditions of the 1930s most intelligent men had bigger and more urgent things to occupy them than the complaints of architectural Purists – the world was too conspicuously going to the dogs in other fields. But after the Second World War, in which a whole generation had been forced to familiarise themselves with machinery on its own terms, the disparity between the observable facts and the architects' Machine Aesthetic had become too obtrusive to be ignored. In the Jet Age these ideas of the 1920s began to wear a very quaint and half-timbered look.

This, of course, made it easier for some feeble intellects to 'adopt a modern style', and we are all familiar with the dandified figures in their draughty and obsolescent sports cars who practise modern architecture as if it were a finished period style with all the answers in the books.

From 'Machine Aesthetic', in Architectural Review *(London), April 1955, pp 225-228. Reprinted with the permission of the author.*

Henry Dreyfuss   1955

Design and Good Living

During the 1940s, product design tended to be not very much more than manipulation of sheet steel, even if its protagonists denied it. Henry Dreyfuss's innovation was to turn it into one of the human sciences. What Dreyfuss did was to replace designers' abstract vapourisings about human aspects of industrial design with the science of anthropometry which he did much to perfect. Throughout his book, Designing for People *(1955), Dreyfuss insists on the importance of taking into account human performance and human dimensions. The title of the book gives away the author's interests; compared with the books of Norman Bel Geddes and Raymond Loewy, Dreyfuss's text is a highly sophisticated one, although this might be expected of one which was also the latest to appear among the publications by the first generation of great industrial designers.*

Like Van Doren, Henry Dreyfuss uses his book to describe his practice and his techniques. Like many of the other leading industrial designers, his origins in the sales and marketing side of industrial production are evident throughout the book. The text projects an almost classical consumerist optimism about the future which was only possible in the days before energy resources were realised to be finite. Although it is the sub-science of salesmanship which continuously influences Dreyfuss's text, he repeatedly attempts to glamourise his business by stressing the importance of research and field-work in design.

Even after Dreyfuss took his own life, his office continued along the lines he dictated for it. It still handles major accounts, including John Deere tractors and American Airlines, and the firm's interest in anthropometrics has been maintained in the publication of Humanscale, *a major study of anthropometry, which is becoming a standard work of reference in the design field.*

Profile of an Organisation

What makes an industrial design organisation tick? How does it operate?

These are fair questions and frequently asked of the independent designer. But, like many fair questions, they are difficult to generalise upon. The reason is that there is no universal *modus operandi* among designers. Walter Dorwin Teague, Raymond Loewy, Harold Van Doren, David Chapman, Hunt Lewis, and Jean Reinecke are all successful practitioners in the field, yet their organisations range in size from one man to hundreds; their methods are as varied as their personalities. Thus, when I answer these questions by describing my own organisation it is not that it is typical, but that it is the only one on which I am qualified to speak.

I opened my first industrial design office in 1929. It was the time of the stock-market crash and the most catastrophic business and financial collapse in history. My new venture, an unknown and experimental profession headed by an unknown and experimenting designer, seemed to have little chance to survive. It would have been prudent, perhaps, to close the office and let the storm pass. But there was also the long-shot possibility that such a 'depression baby' might make it. Industry was in trouble, and businessmen, caught in a hectic scramble to sell their static merchandise, might come to us for design advice.

The day I opened the office, I felt the urge to make some commemorative gesture. I bought a small potted plant for 25 cents. The plant originally had two pathetic green leaves, but it turned out to be a living lucky piece. Twenty-five years and several moves later, it stands today in our office, luxuriant and ceiling high. In a limited circle, it has

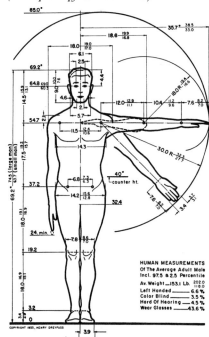

Joe and Josephine, anthropometric sketches by Henry Dreyfuss, 1955 (Henry Dreyfuss Associates)

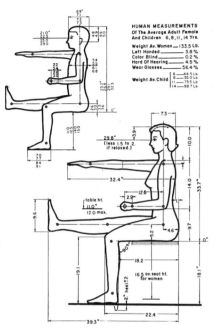

HUMAN MEASUREMENTS
Of The Average Adult Female
And Children 6,8,11,14 Yrs.

Weight Av. Woman —— 133.5 Lb.
Left Handed —————— 3.8 %
Color Blind —————— 0.2 %
Hard Of Hearing ——— 4.5 %
Wear Glasses ———— 56.4 %

Weight Av. Child

6	44.5 Lb
8	55.0 Lb
11	75.5 Lb
14	98.7 Lb

become rather famous. As often as not, old clients who have watched its growth through the years first go over to check its health, then get down to the business that brought them to the office.

In the early days we took what accounts we could get – glass containers, hardware, keys, flower-show exhibits, cedar chests, department store window displays, bottles, hinges, children's furniture, a line of pianos.

I wrote countless letters to big men in industry, hoping to indoctrinate them into the mysteries and advantages of industrial design. Looking back, I think the direct, personal approach of these letters, which explained our work, assured the recipients we could improve their products and increase their business, and requested an interview, was of great value. We have been retained to this day by some of the friends we first contacted by these letters. A New York Central vice-president was asked by a magazine writer how he happened to employ us. He said : 'All I can recall is that at the absolute bottom of the Depression, when thousands of perfectly good railway cars were standing idle all over the country, this youngster followed up a sales letter by walking in with a mess of sketches under his arm and talked us into letting him redesign a bunch of cars and then into designing a whole new Twentieth Century Limited.'

From the outset, I determined to keep our staff small and compact, so we might render a personal service to our clients. As a result, we have restricted our client list to approximately 15, sometimes less, depending on the magnitude of the jobs. If this figure seems small, it should not be interpreted to mean that we have time on our hands. On the contrary, our offices in New York and California are beehives. The explanation is that some of our clients may each produce more than 100 products a

year or request us to oversee the design of products made in as many as 20 factories or may have us designing a fleet of transport planes, simultaneously.

Our goal has always been to become a member of the client's 'family', remaining in touch with his problems, co-operating closely on his current merchandise, but also keeping a sharp eye out for future programs. We feel that we must be faces and personalities – not merely a voice on the telephone, the signature on a letter, or an initial on a drawing.

Six partners head up the organisation. They are supplemented by a hand-picked staff of architects, engineers, designers, artists and sculptors – each doing the work for which he or she is trained and best suited. Some of these co-workers have been with us for more than 20 years. We also maintain a proportionately large clerical staff, which aids in research, scheduling and office routine, so that the creative group may be relieved of . . . this burden.

We work in close collaboration from the generic design to the finished product. The meeting of schedules, development of designs, and the carrying-through process into actual production is the responsibility of the individual partners. They are empowered to hire whatever personnel they need and are in full command of the jobs under their control. On some jobs there is frequent interchange of personnel – members of our staff may work in the client's engineering department, and conversely, we often 'borrow' engineers from a client's staff. When we were working on an airplane, we borrowed an engineer from our aeronautical client to help us over some technical hurdles, and it was almost two years later that we realised we'd forgotten to return him!

Geography determines which jobs will be handled by our New York and California offices. Both offices are available to those clients who have factories near both coasts. A careful system of reports on meetings, phone calls, letters and, of course, progress of current designs keeps both offices fully acquainted with details of all jobs. This arrangement leaves me free to 'float' between the two offices, and my time is about evenly divided between them. At times I almost live by the airplane schedules.

Frequently, each of the partners is in a different city at a different client's factory on a given day. Or we may all appear at Tulsa, Oklahoma, to visit the oil-industry exhibit on a single day. One of our secretaries totalled the office travel for a year and determined that we had flown over a quarter of a million miles.

No matter where I am, I am kept up to date on all work, so that, if necessary, I may step into any job at any time at any place. I receive a large envelope daily – air mail, special delivery – full of vital statistics: correspondence, memos, reports, drawings, photographs of models, blueprints and samples of materials.

We subscribe to scores of publications, which are available to all staff members. Like our research trips through department stores, they help keep us abreast of the times, not only in this country but abroad. Any facts we encounter in these publications which might relate to a client's business are promptly sent to him.

We search incessantly for new ideas, new methods, new materials. We are not content to stand still or to accept present processes as the final word. If we are to progress, we must constantly evaluate the status quo, and if what it stands for is no longer valid, we must abandon it. We look for new points of view, new perspectives. We have found that the creative process is stimulated by new experience and new knowledge.

We have no pat phrase for what makes our organisation tick. Perhaps the secret, if it is a secret, lies in a sincere interest in what we are doing and persistence to an end. To put it another way, our greatest incentive is our desire to create the best . . . designs and then see them get on the market . . .

An Appraisal

In less than half a century it will be AD 2000. Who can say what life will be like then? One can only speculate, knowing that for all the incredible scientific progress of the last 50 years, limitless vistas lie ahead. Perhaps, by the end of the second half of the twentieth century, the one remaining adventure of modern times, travel in space, will have been accomplished. Possibly, the giant eye of Palomar or an even more far-sighted cyclops will have determined whether life exists on neighbouring planets. Conceivably, the Martians will have lived up to their mythological name and attacked the earth and thus united the people on this uneasy sphere . . .

Americans are healthier and more comfortable than ever before. The sweep of technology has not been too much for them. On the contrary, they take for granted, even consider indispensable, things that to their grandfathers could have been little more than a pipe dream – the coast-to-coast dial telephone, radio, television, automobiles, motion pictures, mechanical refrigeration, frozen foods, automatically controlled cooking and washing, Cellophane, color photography, sound amplification, stain-repellent fabrics, and the countless applications of electricity. They casually accept the fact that an airplane travels 1500 miles an hour when they read it in the paper and then go on

RCA television receiver, designed by Henry Dreyfuss, c1955 (Henry Dreyfuss Associates)

'The most widely used device in modern living', Henry Dreyfuss's 'classic' Bell telephone, c1955 (Henry Dreyfuss Associates)

to the comic page to see how Buck Rogers is doing.

This period of mid-century is known as the atomic age and the electronics era. Atomic-powered submarines are already a reality. Ironically, guided missiles, originally conceived for offensive warfare, have become this country's guardian angels. On a more comforting note, research men are working to adapt the products of atomic energy to medicine and to countless every-day uses.

It is part of the industrial designer's business to let his thoughts roam – actually, clients pay him to dream a little – and it is not too difficult to look ahead to the beginning of the twenty-first century when, doubtless, today's sleek designs will seem quaintly amusing to our great-grandchildren.

By then, it is not unreasonable to assume, babies will be immunised at birth against disease, autos or their equivalent will run on atomic power. A great deal of farming will be done chemically, and farm implements will be operated automatically. I have already seen a tractor that can be set to travel in a series of diminishing squares until an entire field is ploughed. No driver is needed, as a special finder wheel is placed in the furrow it made the first time around, and this gradually narrows down, towards the centre. With such equipment a farmer could cultivate half a dozen fields at one time.

It is possible that telephones will have miniature television tubes, so that the person at the other end can be seen, perhaps holding a diagram that is pertinent to the conversation or showing off a newborn babe for Grandpa's examination, back in Iowa.

Mail will probably be dispatched across the country by guided rockets, a civilian application of the guided missile. By this means a letter could span the continent in half an hour.

Rocket oil-well drilling is not improbable. Experiments have already been made, and the rockets, travelling at terrific speed, burn a smooth hole into the earth, making an iron casing unnecessary.

Changes are inevitable in city planning because of traffic congestion, and it's quite possible that in the future people will park their cars on the out-skirts and use fast helicopter transportation to get into downtown sections and crowded areas. Certainly, we have seen only the beginning of helicopters. In the realm of airplane travel, the Pittsburgh airport led the way with a 60-room hotel, including conference rooms. By appointment, people fly in from neighbouring cities, confer with Pittsburgh business associates, then fly off, without ever having left the airport. Similar accommodations have been provided at La Guardia in New York and at Los Angeles International Airport, and more are being built throughout the country.

Wireless transmission of power is possible today, but is not economic. Atomic energy may provide cheap enough power to make it feasible.

The telephone and radio may come into an entirely new development. Just as we can sit in a pool of directed light with the rest of the room in darkness, it is conceivable that one day we will sit in an invisible cone of directed sound wherein we will be able to hear and speak to someone at a distance, as in a phone conversation, but people near us will be excluded from the conversation or, if it is a radio, from the programme.

It is conceivable that all clothing will be made of synthetics, and people will look back on those who used to wear natural skins or fibres as barbarians. The expensive procedures of getting wool from sheep, growing cotton, or having silkworms spin for us may be antiquated. And instead of being sewn together, which leaves uncomfortable seams,

Sarge Steel Special Agent comic illustration (Design Council)

garments may be blended with heat.

The use of microfilm will probably be extended to the point that it will be possible to have the dictionary in a box the size of the telephone, with a lens and spinning device, so that time will be saved in looking up words. The Bible, novels, textbooks, and valuable file material will be made similarly accessible next to your easy chair.

Perhaps cameras will be built into television receivers, so that by pushing a lever a person will be able to obtain a permanent record, which could be developed immediately, the way Polaroid cameras operate today.

What will become of the urge to travel when Paris and New York are only two hours apart by plane? Perhaps we must project ourselves into an ectoplasmic future, when the act of being somewhere will entail no travel at all. A man in Istanbul may decide he wants to spend the evening with friends in Ames, Iowa. The turn of a switch will establish a connection between them and their life-size images in colour and 3-D and their natural voices will be in each other's front parlours. Face-to-face business conferences may be carried on between men whose physical presence is actually in such widely separated places as Rio, Copenhagen and Hong Kong. Remember, 75 years ago, people were not visiting on the telephone or flying through the air or seeing a performer in Hollywood miraculously appearing instaneously on a glass tube in their living room.

Meanwhile, there's disquiet in mid-century, despite the good living that is available to everyone and the prospect of even more. From the headlines and the voices of the commentators and the analyses of the experts comes the suspicion that, without warning, the roof may cave in. From experience, people know that it takes only an arrogant gesture by an ambitious man to start the bombs dropping. A man with a good ear can also detect dissatisfaction of a much milder sort. It is the frequently heard contention that the nation's educators, philosophers, psychologists, and social workers have failed to keep pace with technology in this country. Indeed, some persons argue that the pressing technology of the times is in itself responsible for standardising life, reducing it to the level of uninspired gadgetry, and thus creating an emptiness in the lives of many Americans.

The combination of science, engineering, industrial design and manufacturing skill has brought Americans not only undreamed-of comforts, but a gift of an estimated 1000 more hours of leisure each year than their grandfathers had.

Expressed in another way, the real triumph of the American way of life is that, although the work week has steadily shrunk, productivity has steadily risen . . .

Inevitably the question comes: to what use is this leisure being put? Is it making its recipients happier, better equipped to live a full life, to realise their full potential of personal development? The positive gains are obvious – physical well-being, higher level of health, a life span prolonged nine years in the last quarter century. But does leisure produce happiness? We can scarcely claim to have generated a happier people when we see mounting tension all around us, when it is estimated that one in 20 Americans will spend part of his life in a mental hospital. Nor can we be smug about the blessings of leisure when we consider that many industrial employees who reach the age of retirement plead to be permitted to continue working rather than accept a pension. They don't know what else to do with their time. Have we created leisure only to train a nation of passive participants, filling their time with wrestling on TV, comic books and predigested digests?

Certainly, if the men who make our mass-produced conveniences are to blame for leading people into spiritual frustration, the industrial designer must share the blame. But I do not concede that Americans are overwhelmed by materialism because they enjoy good living. The figures show that there is an accompanying surge of interest, even enthusiasm, in the fine arts, which can be attributed at least in part to the improved design of the ordinary subjects that are all around them. I sometimes wonder if those who put the question really want an answer or are more interested in keeping alive a quarrel.

From Designing for People *Simon & Schuster, New York, 1955, pp 226-240. Reprinted with the permission of Mr John Dreyfuss, executor of the Henry Dreyfuss estate.*

Cars are our Cathedrals

The Citroën DS19, of all automobiles, was the cause célèbre of industrial designers everywhere. Besides employing highly advanced technology, the big Citroën (which, because of the phonetic accident of the type-number when pronounced in French, became known as the Goddess) had a visual integrity which no predecessor had ever rivalled. The public marvelled at the smoothness and drama of its appearance, and in this famous, brief essay, the French savant Roland Barthes (born in 1915) celebrated this car not as a piece of awe-inspiring machinery to be admired from afar, but as a beautiful expression of the modern will. It is remarkable that at exactly the same time as Barthes elevated a mass-produced vehicle to the status of a cult object in France, artists and writers associated with the Independent Group in London were beginning also to make the public aware of the totemic quality of everyday imagery which, translated, became what we now call Pop Art. By the later 1950s machines were no longer being held up as exemplars for artists, they were considered in their own right.

I think that cars today are almost the exact equivalent of the great Gothic cathedrals: I mean the supreme creation of an era, conceived with passion by unknown artists, and consumed in image if not in usage by a whole population which appropriates them as a purely magical object.

It is obvious that the new Citroën has fallen from the sky inasmuch as it appears at first sight as a superlative *object*. We must not forget that an object is the best messenger of a world above that of nature: one can easily see in an object at once a perfection and an absence of origin, a closure and a brilliance, a transformation of life into matter (matter is much more magical than life), and in a word a *silence* which belongs to the realm of fairy-tales. The DS – the 'Goddess' – has all the features (or at least the public is unanimous in attributing them to it at first sight) of one of those objects from another universe which have supplied fuel for the neo-mania of the eighteenth century and that of our own science-fiction: the *Déesse* is *first and foremost* a new *Nautilus*.

This is why it excites interest less by its substance than by the junction of its components. It is well known that smoothness is always an attribute of perfection because its opposite reveals a technical and typically human operation of assembling: Christ's robe was seamless, just as the airships of science-fiction are made of unbroken metal. The DS19 has no pretensions about being as smooth as cake-icing, although its general shape is very rounded; yet it is the dovetailing of its sections which interests the public most: one keenly fingers the edges of the windows, one feels along the wide rubber grooves which link the back window to its metal surround. There are in the DS the beginnings

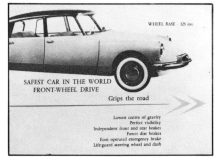

Advertising literature for the Citroën DS19 motor car, c1957 (National Motor Museum)

Still from the film Metropolis, *directed by Fritz Lang, 1926 (Friedrich-Wilhelm-Murnau-Stiftung / Transit Filmgesellschaft mbh, Munich)*

Cover illustration, by Philip Castle, for Mythologies *by Roland Barthes, Paladin, 1973 (Photo Design Council)*

of a new phenomenology of assembling, as if one progressed from a world where elements are welded to a world where they are juxtaposed and held together by sole virtue of their wondrous shape, which of course is meant to prepare one for the idea of a more benign Nature.

As for the material itself, it is certain that it promotes a taste for lightness in its magical sense. There is a return to a certain degree of streamlining, new, however, since it is less bulky, less incisive, more relaxed than that which one found in the first period of this fashion. Speed here is expressed by less aggressive, less athletic signs, as if it were evolving from a primitive to a classical form. This spiritualisation can be seen in the extent, the quality and the material of the glass-work. The *Déesse* is obviously the exaltation of glass, and pressed metal is only a support for it. Here, the glass surfaces are not windows, openings pierced in a dark shell; they are vast walls of air and space, with the curvature, the spread and the brilliance of soap-bubbles, the hard thinness of a substance more entomological than mineral (the Citroën emblem, with its arrows, has in fact become a winged emblem, as if one was proceeding from the category of propulsion to that of spontaneous motion, from that of the engine to that of the organism).

We are therefore dealing here with a humanised art, and it is possible that the *Déesse* marks a change in the mythology of cars. Until now, the ultimate in cars belonged rather to the bestiary of power; here it becomes at once more spiritual and more object-like, and despite some concessions to neomania (such as the empty steering wheel), it is now more *homely*, more attuned to this sublimation of the utensil which one also finds in the design of contemporary household equipment. The dashboard looks more like the working surface of a modern kitchen than the control room of a factory: the slim panes of matt fluted metal, the small levers topped by a white ball, the very simple dials, the very discreteness of the nickel-work, all this signifies a kind of control exercised over motion, which is henceforth conceived as comfort rather than performance. One is obviously turning from an alchemy of speed to a relish in driving.

The public, it seems, has admirably divined the novelty of the themes which are suggested to it. Responding at first to the neologism (a whole publicity campaign had kept it on the alert for years), it tries very quickly to fall back on a behaviour which indicates adjustment and a readiness to use ('*You've got to get used to it*'). In the exhibition halls, the car on show is explored with intense, amorous studiousness: it is the great tactile phase of discovery, the moment when visual wonder is about to receive the reasoned assault of touch (for touch is the most demystifying of all senses, unlike sight, which is the most magical). The bodywork, the lines of union are touched, the upholstery palpated, the seats tried, the doors caressed, the cushions fondled; before the wheel, one pretends to drive with one's whole body. The object here is totally prostituted, appropriated: originating from the heaven of *Metropolis*, the Goddess is in a quarter of an hour mediatised, actualising through this exorcism the very essence of petit-bourgeois advancement.

From 'The New Citroën', originally published in Mythologies *Editions du Seuil, Paris, 1957. From the English translation by Annette Lavers, Jonathan Cape, London, 1972, pp 88-90. Reprinted with the permission of Jonathan Cape Ltd.*

Richard Hamilton 1960

Just what it is

Richard Hamilton (born in 1922) became one of the most articulate of all the British Pop Artists, whose celebration of popular imagery grew out of the 'This is Tomorrow' exhibition at the Institute of Contemporary Arts, where the following extract was originally delivered as a lecture. It was subsequently published in Design magazine, where it entailed a lively debate among readers (including George Nelson and Reyner Banham) who felt their puritanism being impugned by Hamilton's enthusiastic arguments whose political character, if it nowadays seems naive, in 1960 seemed extravagantly heterodox.

Hamilton's paintings have employed mass-produced consumer durables, perhaps Plymouth cars or Braun toasters, as motifs. The original lecture and article drew much of their force from his skilled employment of garish advertising art. With Edgar Kaufmann's solemn investigation of what had happened to design between 1950 and 1960, Hamilton's polemic marks a change in direction for popular debate about the shape of things. Realising the awesome power which the commercial value of beauty and novelty had given industrial designers, Hamilton energetically debunks the fanciful idealistic aesthetics of the 1920s and 1930s (which maintained position during the 1950s as well) and posits instead a short-term, low-rent, chromium Utopia for motivation researchers and make-up artists.

The 1950s have seen many changes in the human situation; not least among them are the new attitudes towards those commodities which affect most directly the individual way of life – consumer goods. It is now accepted that saucepans, refrigerators, cars, vacuum cleaners, suitcases, radios, washing machines – all the paraphernalia of mid-century existence – should be designed by a specialist in the look of things. Of course, the high-power virtuoso industrial designer is not a new phenomenon – Raymond Loewy and Walter Dorwin Teague have been at it for a good many years. William Morris and Walter Gropius realised the potential. What is new is the increased number of exponents, their power and influence upon our economic and cultural life. Design is established and training for the profession is widespread.

The student designer is taught to respect his job, to be interested in the form of the object for its own sake as a solution to given engineering and design problems – but he must soon learn that in the wider context of an industrial economy this is a reversal of the real values of present-day society. Arthur Drexler has said of the automobile: 'Not only is its appearance and its usefulness unimportant . . . What is important is to sustain production and consumption.' The conclusion that he draws from this is that 'if an industrialised economy values the process by which things are made more than it values the thing, the designer ought to have the training and inclinations of a psychoanalyst. Failing this he ought, at least, to have the instincts of a reporter, or, more useful, of an editor.'

The image of the 1950s shown here is the image familiar to readers of the glossy magazines – 'America entering the age of everyday elegance';

the image of *Life* and *Look*, *Esquire* and the *New Yorker*; the image of the 1950s as it was known and moulded by the most successful editors and publicists of the era, and the adman who sustained them – 'the fabulous fifties' as *Look* has named them. Being 'plush at popular prices' is a prerogative that awaits us all. Whether we like it or not, the designed image of our present society is being realised now in the pages of the American glossies by people who can do it best – those who have the skill and imagination to create the image that sells and the wit to respond humanly to their own achievements.

The present situation has not arrived without some pangs of conscience. Many designers have fought against the values which are the only ones that seem to work to the economic good of the American population. There is still a hangover from the fortyish regret that things do not measure up to the aesthetic standards of pure design; the kind of attitude expressed in 1947 by George Nelson when he wrote: 'I marvel at the extent of the knowledge needed to design, say, the Buick or the new Hudson – but I am also struck by my inability to get the slightest pleasure out of the result.' There has since been a change of heart on both sides; on the part of the designers, the men who establish the visual criteria, towards a new respect for the ability of big business to raise living standards – and an appreciation, by big business, of the part that design has to play in sales promotion. What was new and unique about the 1950s was a willingness to accept a new situation and to custom build the standards for it.

There is not, of course, a general acceptance of this point of view. Some designers, especially on this side of the Atlantic, hold on to their old values and are prepared to walk backwards to do so.

Misha Black goes so far as to suggest that advanced design is incompatible with quantity production when he says: 'If the designer's inclination is to produce forward-looking designs, ahead of their acceptability by large numbers of people, then he must be content to work for those manufacturers whose economic production quantities are relatively small.' While Professor Black was consoling the rearguard for being too advanced Lawrence Alloway was stressing the fact that 'Every person who works for the public in a creative manner is face to face with the problem of a mass society.' It is just this coming to terms with a mass society which has been the aim and the achievement of industrial design in America. The task of orientation towards a mass society required a rethink of what was, so convincingly, an ideal formula. Function is a rational yardstick and when it was realised in the 1920s that all designed objects could be measured by it, everyone felt not only artistic but right and good. The trouble is that consumer goods function in many ways; looked at from the point of view of the business man, design has one function – to increase sales. If a design for industry does not sell in the quantities for which it was designed to be manufactured then it is not functioning properly.

The element in the American attitude to production which worries the European most is the cheerful acceptance of obsolescence; American society is committed to a rapid quest for mass mechanised luxury because this way of life satisfies the needs of American industrial economy. By the early 1950s it had become clear in America that production was no problem. The difficulty lay in consuming at the rate which suited production and this rate is not only high – it must accelerate. The philosophy of obsolescence, involving as it does the creation of short-term solutions, designs that do not last, has had its drawbacks for the designer – the moralities of the craftsman just do not fit when the product's greatest virtue is impermanence. But some designers have been able to see in obsolescence a useful tool for raising living standards. George Nelson in his book *Problems of Design*, states the case very forcibly: 'Obsolescence as a process is wealth producing, not wasteful. It leads to constant renewal of the industrial establishment at higher and higher levels, and it provides a way of getting a maximum of good to a maximum of people.' His conclusion is: 'What we need is more obsolescence not less.' Mr Nelson's forward-looking attitude squarely faced the fact that design must function in industry to assist rapid technological development; we know that this can be done by designing for high production rates of goods that will require to be renewed at frequent intervals.

The responsibility of maintaining the desire to consume, which alone permits high production rates, is a heavy one and industry has been cross-checking. With a view to the logical operation of design, American business utilised techniques which were intended to secure the stability of its production. In the late 1940s and early 1950s an effort was made, through market research, to ensure that sales expected of a given product would, in fact, be available to it. Months of interrogation by an army of researchers formed the basis for the design of the Edsel, a project which involved the largest investment of capital made by American business in post-war years. This was not prompted by a spirit of adventure – rather it was an example of the extreme conservatism of American business at the time. It was not looking to the designer for inspiration but to the public, seeking for a composite image in the hope that this would mean pre-acceptance in gratitude for wish-fulfilment. American business simply wanted the dead cert. It came as something of a shock when the dead cert came home last. The Edsel proved that it took so long to plan and produce an automobile that it was no good asking the customer what he wanted – the customer was not the same person by the time the car was available. Industry needed something more than a promise of purchase – it needed an accurate prophecy about purchasers of the future. Motivation research, by a deeper probe into the sub-conscious of possible consumers, prepared itself to give the answer.

It had been realised that the dynamic of industrial production was creating an equal dynamic in the consumer, for there is no ideal in design, no pre-determined consumer, only a market in a constant state of flux. Every new product and every new marketing technique affects the continually modifying situation. For example, it has long been understood that the status aspect of car purchasing is of fundamental importance to production. Maintaining status requires constant renewal of the goods that bestow it. As *Industrial Design* has said: 'post-war values were made manifest in chrome and steel'. But the widespread realisation of aspirations has meant that gratification through automobile ownership has become less effective. Other outlets, home ownership and the greater differentiation possible through furnishing and domestic appliances have taken on more significance. Company policy has to take many such factors into account. Decisions about the relationship of a company to society as a whole often do more to form the image than the creative talent of the individual designer. Each of the big manufacturers has a design staff capable of turning out hundreds of designs every year covering many possible solutions. Design is now a selective

Ford Edsel Citation motor car, 1958. A people's Cadillac which was born for success via market research, but which hit the market cold and soon folded (National Motor Museum)

process, the goods that go into production being those that motivation research suggests the consumer will want.

Most of the major producers in America now find it necessary to employ a motivation staff and many employ outside consultants in addition. The design consultants of America have also had to comply with the trend to motivation research and *Industrial Design* reports that most now have their own research staffs. This direction of design by consumer research has led many designers to complain of the limitation of their contribution. The designer cannot see himself just as a cog in the machine which turns consumer motivations into form – he feels that he is a creative artist. Aaron Fleischmann last year, in the same *Industrial Design* article, expressed these doubts: 'In the final analysis, however, the designer has to fall back on his own creative insights in order to create products that work best for the consumer; for it is an axiom of professional experience that the consumer cannot design – he can only accept or reject.' His attitude underrates the creative power of the yes/no decision. It pre-supposes the need to reserve the formative binary response to a single individual instead of a corporate society. But certainly it is worthwhile to consider the possibility that the individual and trained response may be the speediest and most efficient technique.

Design in the 1950s has been dominated by consumer research. A decade of mass psychoanalysis has shown that, while society as a whole displays many of the symptoms of individual case histories, analysis of which makes it possible to make shrewd deductions about the response of large groups of people to an image, the researcher is no more capable of creating the image than the consumer. The mass arts, or pop arts, are not popular arts in the old

sense of art arising from the masses. They stem from a professional group with a highly developed cultural sensibility. As in any art, the most valued products will be those which emerge from a strong personal conviction and these are often the products which succeed in a competitive market. During the last ten years market and motivation research have been the most vital influence on leading industrialists' approach to design. They have gone to research for the answers rather than to the designer – his role, in this period, has been a submissive one, obscuring the creative contribution which he can best make. He has, of course, gained benefits from this research into the consumers' response to images – in package design particularly, techniques of perception study are of fundamental importance. But a more efficient collaboration between design and research is necessary. The most important function of motivation studies may be in aiding control of motivations – to use the discoveries of motivation research to promote acceptance of a product when the principles and sentiments have been developed by the designer. Industry needs greater control of the consumer – a capitalist society needs this as much as a Marxist society. The emphasis of the last ten years on giving the consumer what he thinks he wants is a ludicrous exaggeration of democracy; propaganda techniques could be exploited more systematically by industry to mould the consumer to its own needs.

This is not a new concept. Consumer requirements and desires – the consumer's image of himself – are being modified continually now, the machinery of motivation control is already established. At present this control operates through the intuitions of advertising men, editors of opinion and taste-forming mass circulation magazines, and

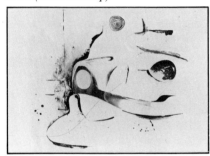

Study for 'Hommage à Chrysler Corporation' by Richard Hamilton, 1957 (Tate Gallery)

the journalists who feed them. But these techniques are too haphazard, too uncertain – fashion is subject to whims and divergencies, to personal eccentricities which squander the means of control. As monopolistic tendencies increase we can expect a more systematic application of control techniques with greater power to instil the craving to consume. It will take longer to breed desire for possession when the objects to be possessed have sprung not directly from the subconscious of the consumer himself, but from the creative consciousness of an artistic sensibility – but the time lag will have distinct advantages for industry.

An industry programmed five years and more ahead of production has to think big and far-out. Product design, probing into future and unknown markets, must be venturesome and, to be certain of success, stylistically and technically valid. As the situation stands at the moment it is anybody's guess (some guessers shrewder than others) which images and symbols will mean most to the public in 1965. It is like someone in 1945 trying to forecast a specific description of Marilyn Monroe. New solutions in product design need to be as inherently likeable and efficient as MM and be as capably presented to the public by star propagandists. Many successful products attain high sales after several years of low production rates. The market is made by the virtues of the object: the Eames chair and the Volkswagen, best-sellers in recent years, are concepts which date back to the 1930s. Detroit cannot wait that long and this impatience is a clue to what we can expect in all the consumer industries. New products need market preparation to close the gap. Industry, and with it the designer, will have to rely increasingly on the media which modify the mass audience – the publicists who not only understand public motivations but who play a large part in directing public response to imagery. They should be the designers' closest allies, perhaps more important in the team than researchers or sales managers. Adman, copywriter and feature editor need to be working together with the designer at the initiation of a programme instead of as a separated group with the task of finding the market for a completed product. The time lag can be used to design a consumer to the product and he can be 'manufactured' during the production span. Then producers should not feel inhibited, need not be disturbed by doubts about the reception their products may have by an audience they do not trust, the consumer can come from the same drawing board.

Within this framework the designer can maintain a respect for the job and himself while satisfying a mass audience; his responsibility to that amorphous body is more important than his estimation of the intrinsic value of the product itself – design has learned this lesson in the 1950s. The next phase should consolidate that understanding of the essential service he is providing for industry and consumer, and extend the use of new psychological techniques as part of the designer's equipment in finding . . . solutions to the needs of society.

From 'Persuading Image', in Design *(London), no 134, February 1960, pp 28-32. Reprinted with the permission of* Design *magazine.*

The Road to Ornament

By 1960 the vision of ever-increasing production and ever-improving conditions was becoming modified. At the same time, the inflexible demands of the Modern Movement that ornament should be abandoned forever were being questioned.

Edgar Kaufmann Jr was one of the leading younger American proselytisers of good design. This extract from Industrial Design *magazine evolved from an appendix which Kaufmann wrote to his* What is Modern Design?, *a book published by the New York Museum of Modern Art in 1950. Kaufmann was cautious about being too prescriptive when it came to defining the conditions of good design, but the public demanded that he did. What this extract represents is ten years of development in the theory of industrial design. The comments of 1960, written after the precepts of 1950, are very revealing. From a too ready acceptance of William Morris's austere codes, industrial design – in the mind of Edgar Kaufmann, at least – was becoming a sophisticated science which paid attention not only to engineering and to art, but also to social awareness. Seen in this light, the way was open for industrial design to assume the position for which it had long been jockeying: to become, in fact, the real art of the twentieth century.*

Design should express the needs and spirit of the times

We still believe in this to a large degree, surely. But in fact the unacknowledged bias has always been toward the more constructive aspects of our needs, of our spirit. Furthermore, the uniqueness of our times seems less important to us than ten years ago, and resemblances to the past or portions of it seem more important. We are willing to dwell on the parallelisms between our needs and feelings and those of other times and places. Our need to be different has become weaker than our need to be ourselves, I'd say.

Design should reflect the leading intellectual concepts and artistic insights of its day

This is a gloss on the above, arising from the speed and spread of communications. But the message to be communicated is changing in character, as noted in part above, and further below.

Structure and materials should express as directly as possible the human requirements that give rise to a design; novel materials and structures should be favoured

Technological novelties continue apace but, with reduced pressure 'to be different', the value of their novelty per se is diminished. Inversely there is a wider, more profound effort to understand human requirements and to express them more adequately and more aptly.

A modern design should not dissimulate its functions, structures or materials

'Honesty of expression' is one of the oldest precepts of modern design, one of the most frequently uttered, and yet less often observed in fact. When Louis Kahn exposes the utilities in his splendid new medical research laboratory building for the University of Pennsylvania, this is no indication that he (or anyone) would do the same in a residence or a concert hall. Who today would shield the motors of a car or a refrigerator with plexiglass? Once, similar ideas were tried. Who would prefer to expose the metal surface of a cast-iron tub? By now it has become clear, though I don't know where it's been said in print, that we expect certain rather effective token elements of practicality to be expressed, serving as a firm basis for more psychological expressions that modulate the design to fit our sense of the appropriate. If this is correctly stated, it means that the road to ornament is once again wide open; and, in fact, ornament is increasingly accepted in modern design today.

A modern design should be integrated as a visually direct and unembellished whole

Part of this is rather like Palladio's *uno intero e ben finito corpo;* the unity of a design is in all ages its passport to existence. A lone kitchen cabinet would reveal to the future something of our sense of style, but rather little of our abilities to organise forms. Yet style, if clearly developed, can imply a whole, as classical remains, for example, testify. So far as 'unembellished' is concerned, it is a dwindling ideal. Much modern ornament is bad, but, after all, subject to improvement. If history is a guide, a tidal ebb and flow is not uncommon between the admiration of simplicity and that of complexity; today the tempo of alternation is a good deal faster than ever before, and that is in fact the mark of our modernity.

Human values should be emphasised in design over technological ones; democratic values, over the desire for elegance or exclusivity

On the whole it seems that these ideas are more powerful than ten years ago. We expect more service and less durability. The special, the rare, retain their status ever more briefly. But it would be an error to suppose this trend irreversible. Technology is evolving made-to-order procedures that, if they catch on and take hold there, will probably affect design in general over the next few decades, as Reyner Banham's '1960' section in the March *Architectural Review* indicates.

One design characteristic not even mentioned in 1950 seems to have resumed a role in 1960 – formality. This is expressed in symmetry, poise, and height that now challenge long-established preferences for occult balance, dynamism, and horizontal thrust. As with simplicity, this seems to be an instance of the tides of taste. Simplicity, to be sure, has another and more profound meaning but one that closely parallels the concepts of unity and integrity, already mentioned. Formality seems to me incapable of such development in depth – it is a surface effect referring to a passing mood, as society is now constituted.

It is possible to draw some tentative conclusions from these shifts concerning the future of design and, in particular, of design education. The conclusions will be no better than the accuracy with which the shifts have been noted, and those who disagree with what has been said so far will be, therefore, even less happy with the rest. First, I believe it's fair to say that design is actively engaged in transforming itself and its relationship to the community. Secondly, these changes, combined with others purely social and technological, make the two leading ideas in design education seem almost equally inadequate. In capsule form, these two ideas are:

(a) a designer should be taught to be competent in his trade, skilled in the technologies (at least the principal ones) of production, presentation and persuasion, and familiar with the design successes and the design ideals of his day.

(b) a designer should be taught to be a scholar and a gentleman, functioning as a professional form-giver who thinks of the community, and its path out of the past into the future, as well as of his own well groomed satisfactions.

I submit that both these formulations (as well as the post-educational one that says a designer is first and foremost a business man) are rooted in a past that is daily slipping away from us. The shift in design ideals – and the shifts of meaning that are eveident even in those precepts that need not be either re-phrased or abandoned – is one piece of evidence to support the claim. The acceptable ideals of design today, and for the closer future, need stating; flexibility and adaptability will be essential, and that means the ideals need to be considered in depth, below the tides of taste and removed from the discolourations of tact. Otherwise there is the prospect that doing will become increasingly thoughtless, and thought increasingly futile, in the field of design.

From 'The Design Shift 1950-1960', in Industrial Design *(New York), vol 7, August 1960, pp 50-51. Copyright 1960* Industrial Design *magazine, Design Publications Inc.*

PRODUCTS

Any selection of photographs presupposes a procedure which existed for their selection. The system used here was a simple, if arbitrary, one. The objects illustrated all have at least two things in common: the first is that they were designed for series production, although not every one achieved this; the second is that they are all exemplars of their type. A trace of the Whig Interpretation sneaks in here because an exemplary illustration supposes that its subject was in the avant-garde of a styling mannerism or a design solution which would not have developed had it not existed.

Generally, the pictures are only of machines. Where chairs or other pieces of inert soft technology are included they are there as outstanding examples of a type, or as control comparisons. Machines are emphasised here, while other surveys of design have included textiles and cabinet-making, glassware and mosaics. The reasons for this are central to the idea of this book; *In Good Shape* is concerned with the appearance of things, the shapes given to complicated machines. Despite the protestations of designers still enthralled by Modern Movement puritanism, most machines do not necessarily have to look beautiful in order to work well. What is interesting here is that it is only when machines begin to assume forms which are not very closely related to the necessities of their operating requirements that the will of the designer can be clearly perceived.

Here is a collection of photographs of pure form. By consideration and comparison of the images, I hope it might be possible to abstract the basic principles which make good design a timeless, but constantly evolving, condition.

Science Museum

Edison Home Model A Phonograph

With Bell's telephone and Marconi's radio, Edison's phonograph was among the first pieces of uniquely twentieth-century equipment with no analogue in any previous age. This is one of the very first examples of its kind and its clumsiness and lack of visual coherence illustrate the principle that when a technology is young there is no aesthetic means of handling it. Only when systems are no longer novel can they consciously be 'designed'.

1900

Ideal Typewriter

Designer Barney & Tanner, New York, USA
Manufacturer Seidel & Naumann, Dresden,
Germany

The Ideal typewriter of Seidel & Naumann dominated the German market at the beginning of the
century and was the prototype of most typewriter
designs for the next 30 years, until the electric
model became commercially possible. It was one of
the first typewriters to have its type-basket disposed at 45 degrees and – although the chassis was
exposed, with no attempt to disguise the mechanicals – by 1900 the Ideal had achieved a level of
design authority which only the new technology of
the 1930s was to replace.

Science Museum

Science Museum

Kodak Brownie Camera

Designer George Eastman
Manufacturer Eastman Kodak, Rochester, New York, USA

The Brownie was intended to supplement the existing and rather more expensive Kodak cameras which Eastman already had available. Like its lineal descendant, the Instamatic of 1963 (see page 211), the Brownie was designed to cope with undemanding conditions and – again like the Instamatic – its design and appearance were uncompromisingly simple. Over 50 million Brownies were manufactured. With these George Eastman created the amateur photography market.

Strowger Telephone

Manufacturer Strowger, Chicago, USA

The Strowger automatic pedestal telephone was typical of the upright type, but was exceptionally cleanly designed in metal. It was not until the 1930s that a standard type of telephone – which is still acceptable today – was established.

Science Museum

Telephone

Manufacturer A/S Elektrisk Bureau, Oslo, Norway

The rococo telephone of 1893 (below), with its absurd decoration and lion's paw feet, had exposed bells. Within eleven years the design was 'rationalised', although the rationalisation involved the deceit of hiding the bells.

Telephone manufactured by A/S Elektrisk Bureau, 1893

A/S Elektrisk Bureau, Oslo   *Norsk Teknisk Museum, Oslo*

1905

Raleigh Bicycle

Designer and manufacturer Raleigh Cycle
Company, Nottingham, England

By 1905 the safety bicycle had reached an established form which can still hardly be bettered in
efficiency. This design goes back to the 1880s,
when the triangulated frame became established as
the best solution to the problem of bicycle design.
The Sturmey-Archer three-speed gears (which
enabled a cyclist to tackle both gradients and the flat
with equal facility) and the Dunlop pneumatic tyres
(which smoothed out any surface) were both contemporary developments which enhanced the utility of the Raleigh bicycle.

Design Council

c1906

Strowger Calling Dial

Manufacturer Strowger, Chicago, USA

The Strowger calling dial achieved a level of visual cleanness and legibility which has never been bettered.

Science Museum

c1908

AEG Electric Kettle

Designer Peter Behrens
Manufacturer AEG, Berlin, Germany

Behrens' nickel-plated steel electric kettle was just one of the many consumer products which he designed while retained by AEG, the massive German electrical products company. All Behrens' designs for AEG, whether they were buildings or kettles, are marked by an austere and elegant simplicity and lack of fuss.

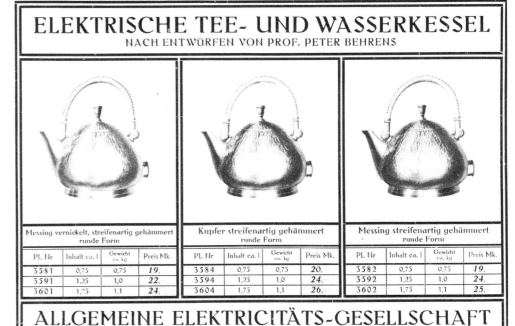

ELEKTRISCHE TEE- UND WASSERKESSEL
NACH ENTWÜRFEN VON PROF. PETER BEHRENS

Messing vernickelt, streifenartig gehämmert runde Form

PL Nr	Inhalt ca. l	Gewicht ca. kg	Preis Mk.
3581	0,75	0,75	19,
3591	1,25	1,0	22,
3601	1,75	1,1	24,

Kupfer streifenartig gehämmert runde Form

PL Nr	Inhalt ca. l	Gewicht ca. kg	Preis Mk.
3584	0,75	0,75	20,
3594	1,25	1,0	24,
3604	1,75	1,1	26,—

Messing streifenartig gehämmert runde Form

PL Nr	Inhalt ca. l	Gewicht ca. kg	Preis Mk.
3582	0,75	0,75	19,
3592	1,25	1,0	24,
3602	1,75	1,1	25,

ALLGEMEINE ELEKTRICITÄTS-GESELLSCHAFT
ABT. HEIZAPPARATE

AEG

Lent to Science Museum, London by EMI Ltd, Hayes, Middlesex

Hymnophon Gramophone

Designer and manufacturer Holzweissig, Leipzig, Germany

This was a disc-playing gramophone. The architectural flourishes such as the marquetry panelling and the free-standing columns belie the fact that this is, in fact, an aesthetically advanced machine. The innovation of concealing the trumpet has allowed one formal concept to dominate the design.

c1911

AEG Electric Fan

Designer Peter Behrens
Manufacturer AEG, Berlin, Germany

Another example of Behrens' product design for AEG.

Collection Manfred Ludewig, Berlin/Photo Jörg P Anders

Die Neue Sammlung, Munich

Motor Boat
MV Ritschy II

Designer M H Bauer

Although not in series production, Bauer's design for a motor boat was featured in the Werkbund *Jahrbuch* for 1912. Because they are often unique designs, boats rarely feature in illustrated books on design, but Bauer's Ritschy II is reproduced here because it is a model of the type of geometrical formalism which dominated the Werkbund imagination in its earliest years.

Leica
Prototype Camera

Designer Oskar Barnack

Oskar Barnack revolutionised photography by producing a high quality, portable 'miniature' camera. The name 'Leica' comes from LEItz CAmera. Barnack was a tireless inventor; this prototype of 1913 appeared 12 years before the first production model, the Model A, which became available to the public with one of Max Berck's Elmar lenses in 1925. The principles which the Leica incorporated then have not been superseded in camera design: in principle, if not perhaps in appearance, the Leica became the iconic camera of the twentieth century.

Michel Auer

Electrolux Model 5 Vacuum Cleaner

Designer and manufacturer Electrolux Aktiebolaget, Stockholm, Sweden

The Model 5 Electrolux vacuum cleaner was the first cylinder, or tank, type to appear on the market. Its form and design were so appropriate for its tasks that they are essentially unchanged today. Besides the convenience of its easy portability, the Electrolux was also popular because it was the first vacuum cleaner that was supplied with various attachments to perform specific cleaning operations. Designed and manufactured originally in Sweden, the Electrolux became so popular in the United States that production started there in 1931.

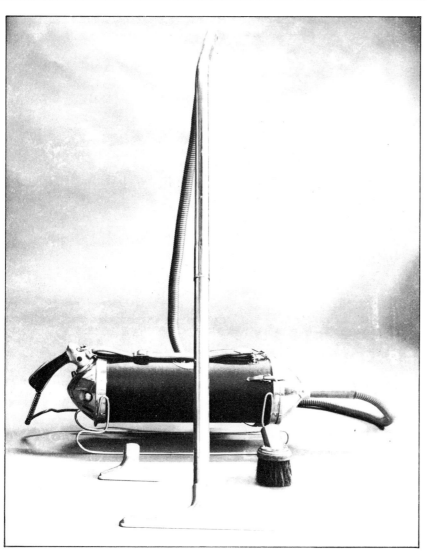

Jay Doblin

Citroën Type A Motor Car

Designer and manufacturer André Citroën, Paris, France

The Citroën Type A, with its disc wheels and slab sides, was Europe's first mass-produced car, and the first to be sold with equipment which up till then had been considered ancillary, that is a hood and a spare wheel, a self-starter and electric lights. It was immensely popular: on its introduction, 16,000 were ordered overnight.

Citroën advertisements from L'Illustration *magazine*

Citroën SA

$3.00 Brings this **Typewriter** on trial!

Easy Payments Like Rent!

Sears, Roebuck

Harris Visible Typewriter

Manufacturer Sears, Roebuck, New York, USA

The Harris Visible typewriter – selling at less than $70 – was typical of the cheap products marketed by Sears, Roebuck. It was neither aesthetically nor technically more advanced than the Ideal typewriter of 1900 (see page 100), although typical of a design solution that had become more or less standardised.

Copper Teapot

Designer Marianne Brandt

Although neither a machine, nor in series production, Marianne Brandt's ebony-handled teapot
is included here, as is Bauer's motor boat (see page
109), as a mature example of a certain sort of style
which became heavily influential. Marianne
Brandt was one of the leading designers to be
associated with the Bauhaus, although she was not,
in fact, an architect. This teapot was made in the
metal workshop of the Bauhaus while it was still in
Weimar.

Germanisches Nationalmuseum, Nürnberg

Victoria and Albert Museum

Chromium-plated Steel Teapot

Designer Naum Slutzky

Like the Marianne Brandt teapot (see opposite), Slutzky's was made in the metal workshop of the Weimar Bauhaus.

Gecophone
Model BC 3200 Radio

Manufacturer The General Electric Company Ltd,
Coventry, England

The Gecophone BC 3200 was a two-valve battery receiver. It is an example of the earliest sort of popular radio, typical of the state of the art when the technology is novel and can be exciting without the application of art. Only when machines such as radios become familiar do they need to be 'designed'.

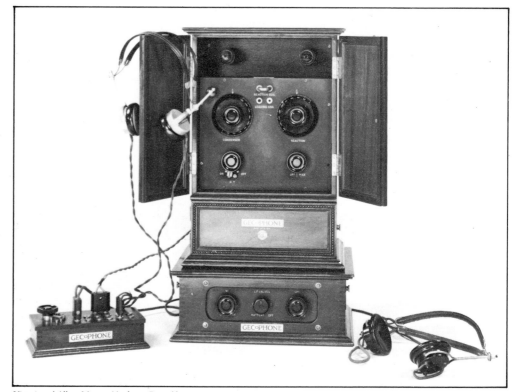

Victoria and Albert Museum/Anthony Constable

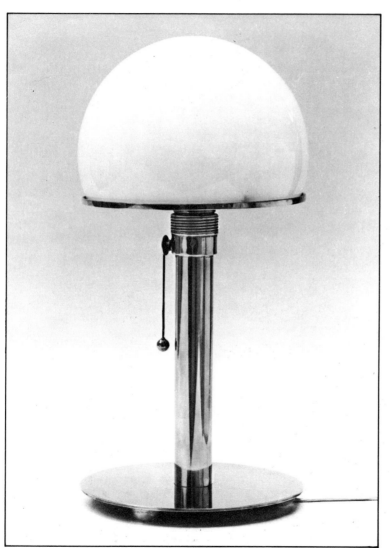

Die Neue Sammlung, Munich

Chromium-plated Steel Table Lamp

Designers Wilhelm Wagenfeld and C J Jucker

Wilhelm Wagenfeld is one of the most prolific designers to have come from the Bauhaus, where this lamp of his was designed and made in 1924. Like the teapots of Brandt and Slutzky (see pages 114 and 115) it is not a machine, nor was it ever mass produced in its own time, but it is illustrated because it is so typical of the design standards which dominated the Weimar Bauhaus. After the Second World War Wagenfeld wrote a book called *Wesen und Gestalt der Dinge um Uns* which described how influential the Bauhaus ideals had been. In fact, a modified version of Wagenfeld's table lamp went on sale in a major chain of British department stores in the late 1970s.

Excelda Gramophone

The Excelda was a miniature disc-playing
machine, typical of the Swiss passion for miniaturisation and precision gimmickry. Its case is a more
or less conscious derivation from that of the Leica
camera (see page 110).

Lent to Science Museum, London by EMI Ltd, Hayes, Middlesex

B3 Wassily Chair

Design Council

Designer Marcel Breuer
Manufacturer Standard-Möbel, Berlin, Germany

Marcel Breuer's Wassily chair, named after the painter Wassily Kandinsky, who was his contemporary at the Weimar Bauhaus, is a classic of modern design, especially if that style label is treated as an historical one. Breuer's intention was to use his chair as an example of how current artistic theories about space could be employed in furniture design.

S33 Tubular Steel Chair

Designer Mart Stam
Manufacturer L & C Arnold, Schorndorf,
Germany

Mart Stam's friends, and the historian Seigfried
Giedion, claim for him the honour of having designed the first cantilevered tubular steel chair, in
Rotterdam in 1926. Stam used these chairs in his
housing at the Weissenhof Siedlung in Stuttgart in
1927. Originally the chair was attributed to Marcel
Breuer and was then known as the B33; nowadays
it is more properly referred to as the S33.

Victoria and Albert Museum

Weissenhof
Tubular Steel Chair

Designer Ludwig Mies van der Rohe
Manufacturer Berliner Metallgewerbe, Berlin, Germany

In the later 1920s, the problem of finding a *parti* for the 'modern' tubular steel chair engaged the imagination of architects and designers in Germany. Mies' example, illustrated here, employed the same principles as Breuer's and Stam's (see pages 119 and 120). Mies' chair was marketed in Germany by Joseph Müller's Berliner Metallgewerbe, which was taken over in 1932 by Thonet, who sold it as the MR534, available with leather or lacquered cane seat and back. (The version without arms, the MR Type APa, was later sold by Thonet as the MR533 and made by Cox in England as their Type M12, also in leather or cane. A very similar design was produced in America by Donald Deskey in the early 1930s.) The canework on the chair shown here was designed by Lilly Reich, who headed the interior design and weaving workshop at the Berlin Bauhaus.

Victoria and Albert Museum

Philips Type 2514
Radio Receiver

Manufacturer Philips NV, Eindhoven,
Netherlands

Like the contemporary Gecophone (see page 116),
this Philips radio was a two–valve battery receiver,
with a two–valve amplifier, but unlike the British
product its astonishing appearance shows that
Philips had made a conscious effort to find a style
appropriate to the new technology.

Victoria and Albert Museum/Rupert Loftus-Brigham

c1927

AEG Hairdrier

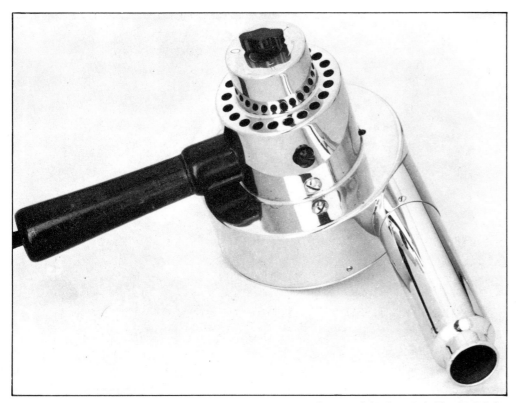

Collection Manfred Ludewig, Berlin/Photo Jörg P Anders

Designed under the supervision of Peter Behrens
Manufacturer AEG, Berlin, Germany

Only the thermoplastic handle suggests that this
hairdrier is not a contemporary machine. The
metal housing (which has been chromium plated)
has no extraneous ornament and possesses a simplicity that is timeless.

Night Light

Designer Marianne Brandt
Manufacturer Körting & Mathiesen AG, Leipzig,
Germany

The Körting & Mathiesen night light was another
example of the designs which the firm produced
under the influence of Bauhaus teaching. This
mode of design has only recently been superseded.

Die Form *magazine/Royal Institute of British Architects, London*

Toledo Counter Scale

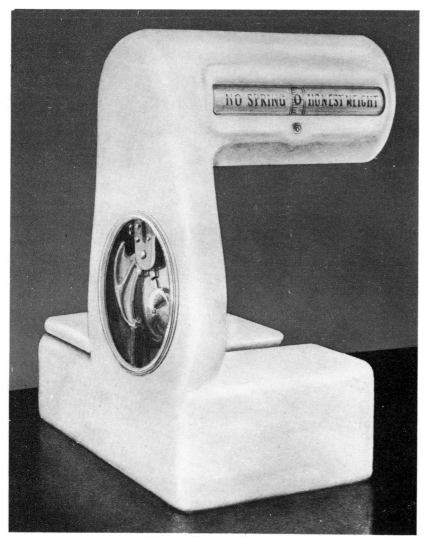

University of Texas and Edith Lutyens Bel Geddes, executrix of the Norman Bel Geddes Estate

Designer Norman Bel Geddes
Manufacturer Toledo Scale Company, Ohio, USA

Counter scales have played an important part in the history of industrial design. Among the first products manufactured by IBM was a scale. The Toledo Scale Company was a market leader. Its first scale, designed in 1897 by Allen de Vilbiss, worked perfectly in engineering terms, but its appearance was accidental (see page 46). When the company hired the designer Norman Bel Geddes to change the appearance of its product, it became one of the first manufacturing organisations in America to be consciously aware of the commercial benefits of sound design.

Siemens Neophone

Manufacturer Siemens Ltd, Brentford, England

This telephone, designed and produced by the British plant of the German electrical company, was the standard GPO type 162 instrument that preceded Jean Heiberg's more visually durable model (see page 134), which had a useful life, at least stylistically, for almost a quarter of a century.

Roger Newport/Sylvia Katz

Gestetner Ltd

Gestetner Model 66 Office Duplicator

Designer Raymond Loewy
Manufacturer Gestetner Ltd, London, England

The Gestetner Model 66 duplicator was Raymond Loewy's first industrial design job, and became the 'before-and-after' image often used to describe the process of product design. It was Sigmund Gestetner himself who gave Loewy the job; in three days and with $3000 Loewy locked himself away, covering the old duplicator with clay so that he could mould a new form for the machine. Although Loewy earnestly stresses the importance of human factors – of ergonomics – in his later designs, the Gestetner job was an exercise in pure formalism. This is not to discredit Loewy's achievement, for what he did to the Gestetner duplicator was to make it into a work of art. The art sold well. Loewy's changed shape successfully rejuvenated the company's image and the Model 66 remained in continuous production until after the Second World War.

Lockheed Sirius
Monoplane

Designer Allan Lockheed
Manufacturer Lockheed Aircraft Corporation,
Burbank, California, USA

The Lockheed Sirius low-wing monoplane never
went into series production in significant numbers,
but it is shown here for two reasons. First of all,
because aesthetically it pioneered some features of
aircraft design which were to become a routine in
the engineer's repertoire and which gained public
currency through the vogue for streamlining. Second, the Sirius is shown here because it was one of
the machines which appealed to the visionary side
of Norman Bel Geddes, who used a picture of the
aeroplane in his famous book, *Horizons*.

*University of Texas and Edith Lutyens Bel Geddes, executrix of the Norman
Bel Geddes Estate*

Lent to Science Musuem, London by EMI Ltd, Hayes, Middlesex

HMV Model 102 Gramophone

Manufacturer The Gramophone Company Ltd, Hayes, Middlesex, England

This illustration of an HMV 102 shows that by the late 1920s, when the technology of the gramophone had become familiar, a standard type of layout had become widely accepted. In this case, the gramophone is disposed in a cabinet rather like a suitcase. This arrangement continued until the later 1950s when the hi-fi boom – and the new technology it entailed – made new shapes and new dispositions of components necessary.

Braun
Radio-Record Player

Designer and manufacturer Max Braun,
Frankfurt, Germany

Braun claims to have been the first manufacturer to
incorporate a radio and record player into a single
unit, thereby creating another opportunity where a
technical innovation had enabled a new design solution to be found.

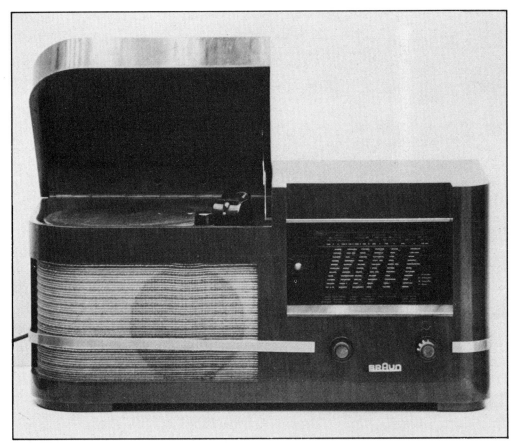

Braun AG

Desk Lamp

Die Form *magazine/Royal Institute of British Architects, London*

Manufacturer Körting & Mathiesen AG, Leipzig, Germany

The manufacturing concern of Körting & Mathiesen actively collaborated with the Bauhaus, which was relatively near to its own plant. The formalist influence of that school can be seen when the desk lamp of 1930 in pastel colours and a very clean shape is compared with the previous model, an altogether visually less clean example of industrial design.

Fuld Telephone

Designer Richard Schadewell
Manufacturer H Fuld, Frankfurt, Germany

The Fuld company had a long history as a telephone manufacturer. Schadewell's design represents one of the many contemporary essays in
establishing a European norm for telephone
design. It is a pleasantly clean aesthetic solution,
but a little more brittle in appearance than
Heiberg's design (see page 134), which was to
become ubiquitous after its introduction in 1931.

Die Form *magazine/Royal Institute of British Architects, London*

Kettle

Wilhelm Wagenfeld

Designer Wilhelm Wagenfeld

Bauhaus formalism is as apparent in this kettle (designed for an electric hot-plate) as it was in Körting & Mathiesen's lamps (see pages 124 and 131). The basic stylistic conceit of employing fundamental geometrical shapes is as apparent here as it was in another product which was influenced by Bauhaus ideals, the Citroën 2CV motor car (see page 155).

Siemens Telephone

Designer Jean Heiberg
Manufacturer Siemens Ltd, Brentford, England

Until 1959 Heiberg's telephone design for Siemens in England remained the GPO's standard instrument. So far had the designer rationalised the visual problems of designing a machine that had to carry out a complex function, yet also be unobtrusive, that his Siemens telephone became an international success, being used as a standard instrument by the Norwegian Elektrisk Bureau, as well as by the British GPO.

Collection Manfred Ludewig, Berlin/Photo Jörg P Anders

1931

Rolleiflex Camera

Michel Auer

Designer and manufacturer **Franke & Heidecke Braunschweig, Germany**

The small twin-lens reflex Rolleiflex was not consciously 'designed', except as a functional and functioning machine, yet – if this idea does not seem to follow Louis Sullivan's too closely and uncritically – the necessities imposed on the manufacturer by the contingencies of easy operation and portability produced a layout that has not been improved upon since. The camera was introduced in 1931 with a Zeiss Tessar lens and was soon imitated by the neighbouring firm of Voigtländer, which produced its Superb camera in 1933. The Rolleiflex is still in production today.

Adler Limousine

Designer Walter Gropius
Manufacturer Adler, Frankfurt, Germany

Gropius' designs for Adler were never mass produced, but are interesting because the Adler was a *Werkbundwagen*, if ever there was one. Aesthetically it represents no advance over Bauer's boat (see page 109) and employs the same design principles as Gropius' office furniture of 1912-14, being essentially a sculptural, geometrical concept rather than a fully thought out solution to the problem of creating a relevant shape for a piece of complex technology. Reyner Banham was the first to point out that, although the Adler is unquestionably a very beautiful motor car, compared with less celebrated vehicles like the Burney Streamliner (see page 54), it was a rather backward piece of design. (Incidentally, however, Gropius' Adler was among the first cars to have reclining seats.)

Die Neue Sammlung, Munich

Zippo Cigarette Lighter

Jay Doblin

Designer and manufacturer George Grant
Blaisdell

The designer of the most popular cigarette lighter
ever made, regulation issue to US forces during the
Second World War, was a non-smoker. Blaisdell
had marketed an Austrian lighter whose design
he admired so much that he began manufacture
himself when the patent expired. His wind-proof
Zippo became one of the most popular of all
mass-produced American artefacts, identifiable in
any part of the world almost as readily as the Coke
bottle (see page 18).

SGE Oriole Stove

Designer Norman Bel Geddes
Manufacturer SGE Company (now defunct)

This was one of Norman Bel Geddes's earliest pieces of industrial design, created just a few years after he opened his studio. He was proud of it and the stove featured prominently in his book *Horizons*. Of it he wrote: 'The first impression is one of utmost simplicity. The stove has no projections or dirt-catching corners, the fewest possible cracks or joints where dirt can accumulate. Burner castings formerly exposed to spillage are protected by an aeration plate which is as easily cleaned as a china bowl.' Bel Geddes's rationalisation might be persuasive, but he entirely forgot to mention – like Loewy with the Gestetner – that, in fact, he also liked the shape.

Preliminary sketches for SGE Oriole stove by Norman Bel Geddes

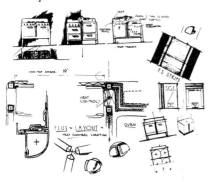

University of Texas and Edith Lutyens Bel Geddes, executrix of the Norman Bel Geddes Estate

Ekco
Model AC74 Radio

Designer Serge Chermayeff
Manufacturer E K Cole & Co Ltd, Southend,
Essex, England

Serge Chermayeff's moulded plastics cabinet for an
Ekco radio was a bold attempt to establish a language of radio design, appropriate to the novelty of
the medium and the advanced technology it
employed.

*Ekco Model BV78 radio with
walnut cabinet, designed
by Misha Black, 1937*

Gordon Bussey   Design Council

139

Huonekalu
Stacking Stools

Designer Alvar Aalto
Manufacturer Huonekalu, Turku, Finland

Alvar Aalto's famous stacking stools were one of
the many experiments made during the 1930s by
architects and designers who wanted to apply
modern materials and techniques to the trad-
itionally craft practice of furniture making. These
stacking stools, which were a familiar feature in
almost every 'modern' interior illustrated in the
heroic books of the age, are still being made today.

Design Council

Braun AG

Braun
Radio-Record Player

Designer and manufacturer Max Braun,
Frankfurt, Germany

After the technical innovation which allowed
Braun to combine in one unit the previously separate components of radios and record players, the
company continued to exploit during the 1930s the
visual opportunities this unified design offered.

1934

Anglepoise Light

Designer G Carwardine
Manufacturer Herbert Terry & Sons Ltd,
Redditch, Worcestershire, England

The young designer G Carwardine approached the
spring manufacturing firm of Herbert Terry at the
beginning of the 1930s with the proposal that they
should build a desk light employing the constant-
tension jointing principles found in the human
arm. The company was persuaded and the
Anglepoise light was the result. It has been in continuous production, scarcely altered other than for
new details and finishes, from that day to this.

Design Council

Chrysler Corporation

National Motor Museum

Chrysler Airflow Motor Car

Designers Carl Breer and the Chrysler Design
Staff
Manufacturer Chrysler Corporation, Detroit,
USA

Walter Chrysler's company was only tenth in the
league of American car sales when the decision was
made to manufacture Carl Breer's revolutionary
Airflow model. The conservative American public
did not take to the unconventional appearance of
the vehicle. Early models had marbled vinyl floor-
covering, moulded roof panels and tripartite bum-
pers, all combined into a streamlined shape that
was almost a decade ahead of its time. The company kept faith with engineer Breer's design but,
even though Norman Bel Geddes was called in to
tidy up the nose (which the public had greeted with
special hostility), the car was still not a commercial
success. It was produced in a number of different
wheelbases under both Chrysler and De Soto
badges, but production ceased in 1937. Imitators
soon followed, however: the Ford Motor Company's Lincoln Zephyr imitated the style and the
image of the Airflow and General Motors derived
their torpedo look – which lasted throughout the
1940s – from it too. It was even said that the shape
of the Airflow influenced Ferdinand Porsche in his
design of the Volkswagen.

Conventional Chrysler sedan, 1934

Citroën 11CV Motor Car

Designer and manufacturer Citroën SA,
Paris, France

Citroën SA

This was the original *traction avant* Citroën, a car whose appearance and engineering were so advanced that even 20 years after its introduction it was ahead of its rivals in both fields. In fact, André Citroën and Henry Ford were close friends, and it is said that the styling of the original 11CV was derived at least in part from Eugene T Gregorie's 1933 designs for English and American Fords. By using front-wheel drive Citroën were able to produce a car which had a wheel at each corner, allowing almost free use of the passenger space aft of the forward-mounted engine. The immensely long wheelbase which this system afforded gave passengers greatly improved ride comfort. Citroën's policy has always been to introduce a radically new motor car and produce it over an extended period. Thus the successor to the 11CV which appeared first in 1934 was the legendary DS19, the Goddess, which first went on sale to the public in 1957 (see page 190).

Electrolux Refrigerator

Royal Institute of British Architects, London/Pencil Points, Reinhold Publishing Corps, 1927

Designer Norman Bel Geddes
Manufacturer Servel Inc (now defunct)

Norman Bel Geddes's Electrolux air-cooled gas refrigerator appeared two years before Raymond Loewy's Coldspot model (see page 147) which he produced for the Sears, Roebuck store. In Bel Geddes's design the circular grey and white nameplate is dominant, but disguises the door catch. The louvres work, but are conceived as ornamental devices rather than necessary engineering features.

Douglas DC-3 Airliner

Designer and manufacturer Douglas Aircraft
Company, St Louis, Missouri, USA

The Douglas DC-3, along with the Boeing 247D
and the original Lockheed Electra, was the first
airliner to make civil aviation commercially realistic. Its smooth, streamlined forms following complex curves attracted gasps of admiration from
both architects and designers who saw in this piece
of high technology a type of beauty, hitherto un-
available to them, which the inevitability of technology had made available to Douglas. The DC-3
was cited as an example worthy of imitation in
influential books by both Le Corbusier and Walter
Dorwin Teague. Douglas had only entered civil
aviation in 1931, when the Ford Motor Company
dropped out. Airlines were looking for an aeroplane to replace the Ford Trimotor, as well as provide in-flight sleeping accommodation. The DC-3,
known as the DST (Douglas Sleeper Transport),
evolved out of the DC-1 and DC-2 series, but was
substantially larger. The DC-3 saw wartime service as the Dakota troop transport and by 1946
10,000 had been built. In 1966, DC-3s still com-
prised almost one third of the world's air transport.

McDonnell Douglas Corporation

Coldspot Refrigerator

Designer Raymond Loewy
Manufacturer Sears, Roebuck, New York, USA

Raymond Loewy was retained by Sears, Roebuck in 1935 and his Coldspot refrigerator is one of the most famous pieces of product design created in this century. With details and techniques borrowed from the automobile industry, its commercial success seemed to endorse Loewy's optimistic interpretation of the utility of his profession. Loewy's new, curved Coldspot (which replaced an already dated angular piece of kitchen equipment) achieved annual sales of 275,000, compared with 60,000 for its predecessor. Loewy's design included inserting mechanical innovations such as his 'feather touch' latch, connected to a foot pedal, which allowed a fully laden housewife still to open the refrigerator. In 1939 Loewy started working for the Frigidaire company.

Raymond Loewy International Inc

Kodak Brownie Camera

Designer Walter Dorwin Teague
Manufacturer Eastman Kodak Company,
Rochester, New York, USA

The Brownie was Eastman's first die-cast camera, and until the arrival of the Instamatic in 1963 established the unchanging iconography of the cheap camera (see page 211). It sold for $1, and by the end of the decade had outsold any other camera ever built. The raised stripes which Teague introduced were – despite their evocation of the streamlining mode – not just for appearance, but were in fact a structural part of the aluminium body intended to reduce the area of the lacquer and thereby obviate cracking.

Walter Dorwin Teague Associates

1936

Pennsylvania Railroad Engine 3768

Designer Raymond Loewy
Manufacturer Pennsylvania Railroad Engineering
Department, Philadelphia, USA

Le Corbusier had spoken of creating new tracks in
the sky, but Loewy applied the aesthetic principles
of aeronautics to railway trains. Loewy said in the
preface to his book, *The Locomotive*, 'My youth
was charmed by the glamour of the locomotive. I
am still under its spell.' Wind-tunnel tests showed
that Loewy's streamlined shape, besides looking
sharp, reduced drag by one third at maximum
speed when compared with a similar engine of
non-streamlined design.

Raymond Loewy International Inc

Service Station
for Texaco Gasoline

Designer Walter Dorwin Teague

Although architecture and corporate identity programmes are not the business of this anthology of
images, Teague's Texaco service station designs
deserve inclusion because they achieved an elemental and archetypal form which was widely
imitated. Teague said of his work for Texaco that
the essential service station 'illustrates the crys-
tallisation of our style'.

Walter Dorwin Teague Associates

1937

Bell Telephone

Henry Dreyfuss Associates

Designed by Bell Laboratories in collaboration
with Henry Dreyfuss
Manufacturer Bell Laboratories

Dreyfuss's Bell Telephone was designed in 1937
and remained in production throughout the 1950s.
It became the most familiar vernacular product of
American industrial design (see page 84). It is the
Dreyfuss organisation's policy to maintain only a
limited number of accounts (see page 82), but these
are all with the biggest corporations. Henry
Dreyfuss Associates retain the Bell account to
this day, but commercial pressures from *arriviste*
rivals have forced them to prepare designs which
pander to less sophisticated public taste than did
Henry Dreyfuss's own timeless solution of 1937.

1937

Minox Camera

Designer Walter Zapp
Manufacturer VEF, Riga, Lithuania

The Minox miniature camera, combining precision engineering with an elegantly simple appearance, soon became a classic. Design commentators have often cited it as a happy example of what the force of necessity – in this case, miniaturisation – does to the appearance of things. The original Minox made 50 exposures; in all its essentials it is unchanged today.

Michel Auer

1938

Murphy
Model A52 Radio

Designer Dick Russell
Manufacturer Murphy Radio Company Ltd;
cabinet made by Gordon Russell Ltd, London,
England

Dick Russell's veneered wooden cabinet for Frank
Murphy's advanced superhet radio of 1938 offers
an interesting contrast to the more radical design
solutions to the problem of the radio cabinet
offered by Serge Chermayeff and Wells Coates (see
pages 139 and 160). Although puritanically free of
extraneous ornament, as one would expect in a
design from the hands of Gordon Russell's brother,
the form and appearance of the cabinet is dependent
on craft work and is, therefore, inappropriate to the
advanced technology used in the receiver itself.

Design Council

1938

Cadillac 60 Special Motor Car

Designer Bill Mitchell
Manufacturer Cadillac Division, General Motors
Corporation, Detroit, Michigan, USA

While still in his twenties, Bill Mitchell became
head of General Motors' Cadillac studio. In his
designs for the prestige division he intended to
imitate the appearance of classic cars such as the
German Mercedes–Benz and the American
Duesenberg and Stutz. Under Mitchell's direction
after the war, Cadillac was to develop an entirely
novel design philosophy all its own.

General Motors

Citroën 2CV
Motor Car

Designer Pierre Boulanger
Manufacturer Citroën SA, Paris, France

The universally loved Deux Chevaux is still in production today. Its design was finalised before the Second World War by Citroën's engineer, Pierre Boulanger, and was, like Porsche's Volkswagen, intended to be a people's car. It was not introduced to the public until 1946 and since then it has sold steadily throughout Europe, being progressively refined within the original specification. The first models appeared with a cyclopean single headlight and were designed to carry 50kg of luggage at 50kph. There is a startling lack of compromise in the design of the Deux Chevaux which, according to Wolfgang Schmittel who referred to the circular door aperture as a telling detail, owes something to Bauhaus design practice.

Citroën 2CV motor car, 1970s

Citroën SA   What Car? *magazine*

Phonola Moulded Plastics Radio

Designers Achille, Pier Giacomo and Livio Castiglioni
Manufacturer FIMI Phonola, Milan, Italy

The Castiglioni brothers' radio cabinet is an astonishingly advanced aesthetic conception. At the same time that British manufacturers were still producing dark, heavy-looking sets, this early-Italian design predicts the light and pastel-coloured appearance of radio receivers throughout the 1950s. Achille Castiglioni, the most famous of the brothers, went on to design light fittings which became classics of modern Italian design.

Stile Industria *magazine*

Ferranti Model 139 Radio

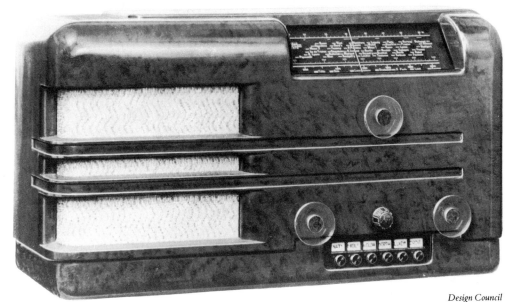

Design Council

Manufacturer Ferranti Ltd, Moston, Lancashire, England

This anonymous Ferranti design of 1939 was among the most advanced of contemporary British radio cabinets. In the highly competitive field of radio cabinet design only three companies, Ferranti, Murphy and Ekco, maintained the highest standards of design throughout the pre-war years.

Parker 51 Fountain Pen

Designers Kenneth Parker, Ivan D Tefft, Marlin Baker and Joseph Platt
Manufacturer Parker Pen Company, Janesville, Wisconsin, USA

The Parker 51 fountain pen, which was produced for the company's fifty-first anniversary, is a classic of modern design in the truest sense, even if that word has been heavily abused in this context before. It has featured both in standard tutelary texts, such as Moholy-Nagy's *Vision in Motion*, and an example is in the collection of the Neue Sammlung in Munich, one of Germany's leading museums of design. The Parker 51 grew out of the company's experiments with fast-drying inks; Parker's own new ink had a high alkali content which tended to rot conventional rubber sacs, so a synthetic material, called Lucite, was employed. Contemporary publicity stressed that the Lucite barrel was designed on ergonomic principles, but, in fact, the Parker 51, elegant aesthetic design solution that it is, relies on established technology dandified by modern dress. Thus it is typical of much American product design. The 51 cost a quarter of a million dollars to develop and by 1970 Parker had spent $20m on its sale and promotion.

S R Gnamm

Raymond Loewy International Inc

International Harvester Farmall Tractor

Designer Raymond Loewy
Manufacturer International Harvester, Chicago,
Illinois, USA

The Farmall tractor was one of Loewy's famous
'before-and-after' transformations. The effect was
that what was once an iron-wheeled contraption,
looking a century old, became clean and modern.
Loewy brought the space-age domestic aesthetic of
the car showroom to the farmyard.

Ekco Model A22 Radio

Designer Wells Coates
Manufacturer E K Cole & Company Ltd,
Southend, Essex, England

Wells Coates continued his work on radio cabinet design after the Second World War. His well known concentric design in moulded plastics with chrome details was selected for the 'Britain Can Make It' exhibition, a prologue to the Festival of Britain of 1951.

Design Council

Vespa Motor Scooter

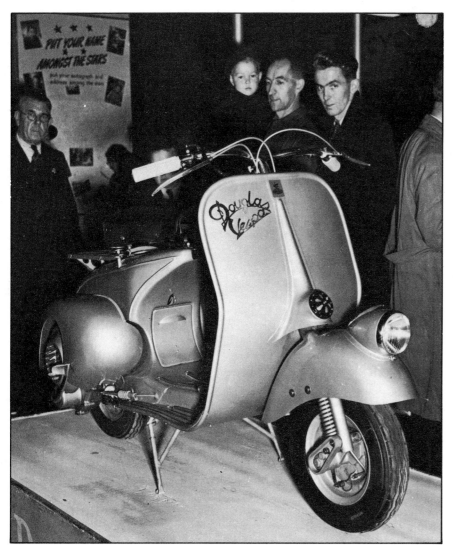

Designer Coradino d'Ascanio
Manufacturer Industrie Aeronautica e Mecanica
Renato Piaggio, Biagno, Italy

The Italian motor scooter was perhaps the only new genus of the species of motorised transport to appear since the Second World War, and Vespa's beautiful clothing of the rude machinery has become an archetype. To assess the visual power of the designer's imagination, the Vespa should be compared with the contemporary Lambretta scooter (inset) which is a far more archaic-looking piece of equipment. The Vespa, which *Design* magazine in its inaugural issue declared to be an example of streamlined design, has remained essentially unaltered for 30 years, apart from the change of position of the headlamp from the mudguard to the handlebars. The early models were manufactured in England under licence by Douglas.

Lambretta motor scooter, 1948

Sport and General   Innocenti, Milan

1948

Olivetti Lexikon 80 Typewriter

Designer Marcello Nizzoli
Manufacturer Ing C Olivetti & C SpA, Ivrea, Italy

Nizzoli's Lexikon 80 typewriter for Olivetti, engineered by Giuseppe Beccio, marks the beginning of that company's self-conscious projection of itself as an organisation which relies on good design to sell its products.

Ing C Olivetti & C SpA/Photo Ballo and Ballo

Supreme in Economy

The new **MORRIS MINOR** makes the most of your petrol, goes farther

on a tankful. Traditional Morris reliability and low maintenance are inherent in this modern design.

National Motor Museum

Morris Minor Motor Car

Designer Alec Issigonis
Manufacturer Morris Motors, Oxford, England

The Morris Minor, when it appeared, was almost as much of a revelation as the Volkswagen had been and it shared many of the Volkswagen's design principles. Issigonis chose unitary construction and a shape that was entirely novel to British eyes, owing nothing to any established precedent. The Morris Minor achieved a higher level of passenger comfort and safety than any pre-war British car and until the appearance of the Mini (see page 204), also designed by Issigonis, its production run was also the biggest. With the development which BMC denied it, the Morris Minor could have rivalled the Volkswagen in international sales and reputation. Issigonis was one of the last engineers working for a major manufacturing concern who was able to maintain a relaxed and informal attitude to his work. There is a story about the final styling of the Morris Minor. When Issigonis saw the completed mock-up he felt the proportions were wrong, so he cut the car in half lengthways and inserted a 10in gusset from front to rear, transforming the car into the shape it retained for its whole life.

Mullard
Model MAS 276 Radio

Manufacturer Mullard Radio Valve Ltd,
Mitcham, Surrey, England

The MAS multi-waveband receiver had very
advanced styling. The Bakelite case allowed a
single visual idea to operate and no extraneous
stylistic details were allowed to obliterate it.

Gordon Bussey

1948-49

Zeiss Contax S Camera

Michel Auer

Designer and manufacturer Zeiss, Dresden, Germany

The Contax S camera was perhaps the very first 35mm camera to employ eye-level reflex viewing. This was made possible by one technological innovation, the pentaprism, and the pattern of small-format reflex design established by Zeiss in 1948 has remained unchanged ever since.

Coffee Machine

Designer Gio Ponti
Manufacturer La Pavoni, Milan, Italy

Although Italian design has achieved a reputation for simplicity and elegance which is, on the whole, entirely justified, it is strange that in his prototype design for this commercial coffee machine (one of the most Italian of all machines) Gio Ponti relies on the idioms of art deco and the American automobile industry.

Gio Ponti

Fiat 1100 Cabriolet Motor Car

Designer Pinin Farina
Manufacturer FIAT SpA (Fabbrica Italiana
Automobile Torino), Turin, Italy

A contemporary critic of this Fiat in *Design* magazine said that it 'combines lightness of appearance with great width and solidity' and if these design criteria seem naive today, at least these qualities were enough to separate completely the little Fiat from even the most advanced American designs. Since the end of the Second World War Italian design has led all others in the field of car styling.

Carrozzeria Pinin Farina SpA/Photo Moncalvo

Olivetti Lettera 22 Portable Typewriter

Designer Marcello Nizzoli
Manufacturer Ing C Olivetti & C SpA, Ivrea, Italy

Marcello Nizzoli's portable typewriter for Olivetti created a visual type which has been continuously imitated since. His design won a Compasso d'Oro award in 1954 and, along with examples of Olivetti's products designed by Ettore Sottsass and others, Nizzoli's Lettera 22 was one of those influential machines which graced Olivetti's dramatic New York showrooms. Here, displayed on pedestals and supplied with rolls of commercial paper, both the sculptural and ergonomic qualities of Olivetti designs were available to influence a whole generation of young American designers. Nizzoli's superbly elemental clothing for the machinery reflects the manufacturing improvement which Giuseppe Beccio incorporated into the Lettera 22: he reduced its number of parts from 3000 to 2000.

Design Council

1950

Saab 92 Motor Car

National Motor Museum

Designer and manufacturer Svenska Aeroplan Aktiebolaget, Linköping, Sweden

The Saab 92 was a car produced by the Swedish Aircraft Company and, not surprisingly, brought the science of streamlining to a new height for a road vehicle. The first 92s had twin-cylinder two-stroke engines, a smooth underpan (to cut down drag) and were said to be capable of 65mph. With a number of changes, most notably in the power plant (which was changed for a German Ford unit in 1966), the Saab 92 design continued in production for 28 years

Braun Model S50
Electric Razor

Designer and manufacturer Max Braun,
Frankfurt, Germany

This is the production version of the prototype electric razor developed by Max Braun in 1938. It was introduced at a trade fair in Frankfurt immediately after the Second World War and since then the system with an oscillator motor and metal foil has been the staple of almost all electric razor design, with the exception of those produced by the Dutch Philips Company. Although the technology was well advanced, the Braun S50 looks visually dated when compared with a contemporary Remington razor (see page 176); by 1953 Braun had caught up with the new style of the Remington. In 1954, Braun did a licence deal with the American Ronson Corporation which ensured the establishment of the company's technology on both sides of the Atlantic.

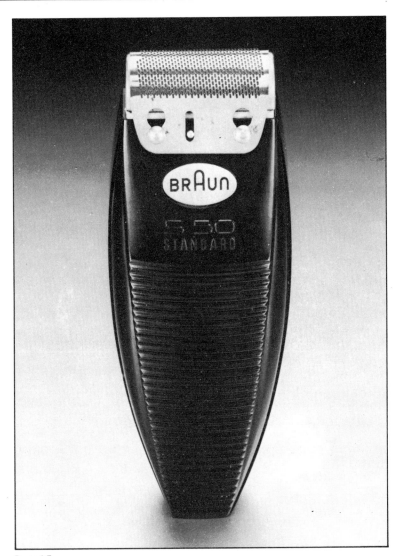

Braun AG

Science Museum

Remington Standard Typewriter

Designer and manufacturer Remington Rand Corporation, Elmira, New York, USA

The Remington Standard, called the 'Super Riter', was a development of a successful line of office model typewriters. The design of the casing, which the manufacturers stressed was there to keep out the dust, shows the influence of Olivetti styling on the work of the big American corporations.

KB Model FB10C Radio

Designer Lawrence Griffin
Manufacturer Kolster Brandes Ltd, Foot's Cray,
Kent, England

This strikingly ugly little radio is a telling example
of how designers and manufacturers struggled in
the early 1950s to get away from the stereotypes of
pre-war radio cabinet design. As in this case, the
designs were not always very happy, the stylist
having picked up some of the less attractive idioms
of American automobile styling and science fiction
films.

Victoria and Albert Museum

Porsche Model 356 Motor Car

Porsche AG

Designer and manufacturer Porsche AG, Stuttgart, Germany

If the Volkswagen was the car that Hitler invited Ferdinand Porsche to design, the Porsche Model 356 was the car the man himself *wanted* to build. Its ancestry in the People's Car is clear, although its relationship with the Mercedes S, SS and SSK (which Porsche also designed) is less apparent. The type number derives from the number of the design study which the firm made since the setting up of its Stuttgart studio in 1931.

The design concept is similar to the Volkswagen, but much more refined. Both cars had unitary construction and a rear-mounted flat four-cylinder engine which was air cooled. Both were highly streamlined. The Model 356, whose detail design was completed by Ferry Porsche Jr, remained in production until 1965. Its bodywork was designed on aerodynamic principles by the happily named Erwin Kommenda.

1952

Chair in Steel Wire

Designer Harry Bertoia
Manufacturer Form International Ltd, London,
England, under licence from Knoll International
Inc, New York, USA

Together with Eames's rosewood chair for
Herman Miller (see page 191), Bertoia's steel wire
lattice design for Knoll International is among
the best known of recent designs. Both featured
continuously in the interiors of architect-designed
houses illustrated during the 1950s and 1960s, in the
same way as Breuer and Stam and Mies chairs had
done in the decade before the Second World War.

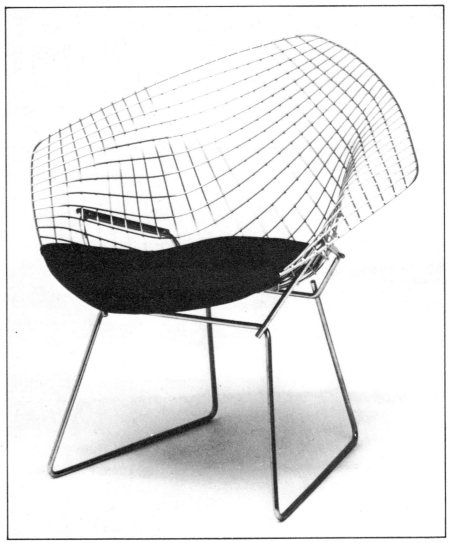

Design Council

National Motor Museum/Jerry Chesebrough

Hillman Minx Mark VIII Californian motor car, 1955

National Motor Museum

Studebaker Commander Motor Car

Designer Raymond Loewy
Manufacturer Studebaker, South Bend, Indiana, USA

Raymond Loewy was hired by Studebaker's Paul Hoffman in 1933. His design of the Champion in 1939 raised sales from $43m to $81m in 1939-40. With his designs for Studebaker cars Raymond Loewy introduced the scale, proportions and details of the best European sports cars to the American manufacturers. This was a complete revelation to the American public, and to Studebaker's rivals, and helped temporarily to pull the ailing company out of a financial hole. In the years after its introduction, Loewy's dramatic Studebaker Starline coupé held 40 per cent of the company's car sales. The Studebaker designs were selected by *Reader's Digest* as exemplary and since then have influenced all subsequent automobile design in the United States. Loewy was also retained by the Hillman division of the Rootes Group in England and, turning the tables again, he brought to the British Isles the style and luxuries of American cars in his designs for the Hillman Minx and its derivative, the Hillman Californian (inset). Although Loewy's Studebaker designs won constant praise from English critics during the 1950s, the company became desperate and Loewy's design was dropped for the 1957 model year in favour of a more traditional transatlantic treatment.

Remington Electric Razor

Designer Alan Irvine
Manufacturer Remington, New York, USA

The Remington electric razor design of 1953, as well as employing different technology from the contemporary Braun (see page 170), was also more advanced stylistically. Its two-piece moulded plastics body and carrying case were all of a piece with current American product design: both Loewy's cars and Irvine's razor share the same curves.

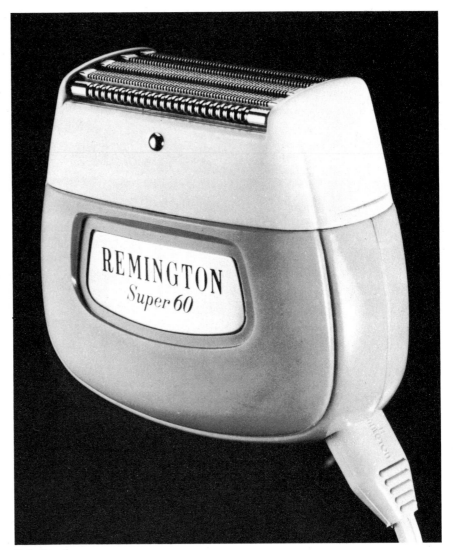

Design Council

Isetta Motor Car

Designer and manufacturer Iso, Bresso, Milan, Italy

Although the Suez crisis of 1956 created a rash of bubble cars, the Isetta three-wheeler preceded them all. Problems in stability led eventually to its growing a fourth wheel at the back. The earliest models, like the Citroën 2CV (see page 155), had canvas tops and all incorporated an ingenious solution to entering such a compact vehicle: the steering column was hinged and moved with the front door to allow easy access. Never a great commercial or dynamic success, the Isetta's bulbous visual unity was in a direct line which was to lead to Dante Giacosa's award-winning Fiat 500 (see page 199).

Gill Streater

1953

Chevrolet Corvette Motor Car

Designer Harley J Earl
Manufacturer Chevrolet Division, General
Motors Corporation, Detroit, Michigan, USA

Harley Earl's Chevrolet Corvette popularised the
wrap-around windscreen, an entirely meretricious
styling device which became a fundamental part of
the vocabulary of automobile design during the
1950s and early 1960s. The Corvette was an American attempt to come to terms with the European
two-seater sports car which had taken over a large
sector of the domestic market. It was produced in
an initial run of 300 in glass fibre, the first production car to employ this material. In everything
except looks the Corvette was a compromise: it
was a limited success at first because traditional
sporty features such as inside door latches and flap-
ping sidescreens put off *boulevardiers*, and the
slushy GM transmission put off the sports car fanatics who had been buying European. For the 1956
model year, when production became established,
Harley Earl made essential changes to his design,
under the influence of the Mercedes-Benz 300SL.

Chevrolet Motor Division

AB Volvo

Volvo Amazon Motor Car

Designer Jan Wilsgaard
Manufacturer Volvo Aktiebolaget, Gothenburg, Sweden

The Volvo Amazon remained in production for 18 years. When it appeared in 1953, its appearance could be said to be neither avant-garde nor *retardataire* but, like the Peugeot 403, it was a clean and decent design solution whose ruggedness and lack of fuss were appropriate to the reputation of the Volvo mechanical parts which it enclosed. Compared with contemporary British cars, such as the Farina-inspired Austin Cambridge (inset), the Volvo has a timeless rightness, if not beauty. The British car has ill conceived, tawdry details and gives a nagging impression of thinness and instability. In contrast, the way the Volvo's body panels follow constant curves and the manner in which it sits four-square on its wheels convey an impression of solidity which its mechanism does not belie.

National Motor Museum

Austin A55 Cambridge motor car, 1957-58

Nash Metropolitan Motor Car

Designer AMC, Southfield, Michigan, USA
Manufacturer Austin Motor Company Ltd,
Birmingham, England

Selling at less than $1500, the Nash Metropolitan
was intended to reverse the pattern which Loewy
had made so successful with his Studebaker model
line. It was intended to meet American standards of
comfort, performance and appearance, but to have
European compactness and economy. It was
designed and engineered in the United States by the
Nash Rambler Company, but made in England by
Austin, using components which were essentially
no different from those used in the contemporary
Austin A35. Even though it was said to be comparable to larger American models in its interior
space and appointments, the Nash Metropolitan
only enjoyed a brief vogue as a curiosity in the
United States and Britain, and was never a success
on either side of the Atlantic.

National Motor Museum

Henry Dreyfuss Associates

Hoover Vacuum Cleaner

Designer Henry Dreyfuss Associates
Manufacturer Hoover Company, North Canton,
Ohio, USA

This compact model was brought in to rival the
temporarily more popular horizontal cylinder vacuum cleaner. The main frame was a single piece
die-casting, with the motor mounted from below
during assembly; a light was included in the casing
so that dark corners could be illuminated during
housework.

1954

Vickers Viscount Airliner

Designer George R Edwards
Manufacturer Vickers Armstrong Ltd,
Weybridge, Surrey, England

When the American Capital Airlines bought 60
Vickers Viscounts a new period in civil aviation
had begun. What the Douglas DC-3 (see page 146)
and DC-4 had been to the later 1930s and 1940s, the
turbo-prop Viscount was to the 1950s and early
1960s. It filled the technology and comfort gap
between the DC-3 and the pure jet Boeing 707 (see
page 187).

British Aerospace

Parker Ball-point Pen

Parker Pen Company

Designers Nolan Rhoades (styling) and Edward
Grumich (interior)
Manufacturer Parker Pen Company, Janesville,
Wisconsin, USA

The established Parker Pen Company held back
from producing a ball-point pen until the technology had been sorted out by more buccaneering
manufacturers. Parker did not want its name
associated with an expendable product; until the
company developed its retractable pen in 1954, the
ball-point idea was associated in the public mind
with cheap pens which had caps and which could
easily be discarded. The Parker product image was
one of quality and longevity so it was necessary, if
the company wanted to get into the lucrative ball-
point market, to develop a pen which would have
these attributes. The Parker retractable nib system
included a nylon ratchet that turned the point every
time it was retracted, thus avoiding the uneven
wear which had bedevilled cheaper ball-point pens.

Ford Thunderbird Motor Car

Designer Bill Burnett
Manufacturer Ford Motor Company, Detroit, Michigan, USA

The T'bird was Ford's sports car, its reply to the European mode which Raymond Loewy had begun for Studebaker and General Motors had continued in their two-seater Corvette (see page 178). The design programme started in 1953-54; the first production model won critical praise for its clean, uncomplicated looks (which were partially transferred to English Ford saloons, such as the Consul), but it did not sell well. The two-seater was replaced in 1958 by the four-place Squarebird, an altogether more American sort of design.

National Motor Museum

1955

Braun Model SK2 Radio

Braun AG

Designers Artur Braun and Fritz Eichler
Manufacturer Max Braun AG, Frankfurt,
Germany

This Braun radio design appeared at just that time
when the company was beginning to develop
under Dr Fritz Eichler its highly self-conscious
design policy which has since given it an international reputation.

Hyster Fork-lift Truck

Designer Henry Dreyfuss Associates
Manufacturer Hyster Corporation, Portland,
Oregon, USA

Henry Dreyfuss Associates has had a long-term
association with the Hyster Corporation. Of his
design for one of its fork-lift trucks he wrote: 'The
designer's temptation was to endow this Hyster lift
truck with a racy appearance, but, after study, it
was decided to give it a look of simple, rugged
functionalism.' However, Dreyfuss's idea of functionalism would have alarmed one of the messianic
proponents of that aesthetic of the 1920s or 1930s:
his Hyster truck is, in fact, an elegant, refined and
subtle work of art whose appearance bears only a
slight relation to its operating necessities.

Henry Dreyfuss Associates/Hyster Company

Boeing 707 Jetliner

Boeing Commercial Airplane Company

Designers Boeing Staff and Walter Dorwin
Teague Associates
Manufacturer Boeing Aircraft Company, Seattle,
Washington, USA

The Boeing 707, which followed the British
Comet, was the first pure jet aircraft to make jet
travel a commercial reality. It has become the
iconic jetliner, although technologically little more
advanced than the earlier Comet. Walter Dorwin
Teague Associates advised on the nose contours
and the exterior markings: the original flying prototype (shown here) was coloured golden yellow
on top, with a silver belly and charcoal-brown
stripes and engine pods. Teague had become
associated with Boeing in his work on the interiors
of the Stratocruisers. Of his work a Boeing
spokesman said it was 'not a talent for dressing-up
products, but the systematic application of creative
thinking'. (The paint job and interior of the United
States' Presidential jet, Air Force 1, a Boeing 707,
was done by Raymond Loewy.)

US Air Force photo by Ken Hackman

US Presidential aeroplane, Air Force 1

Stenafax Duplicator

Designer Henry Dreyfuss Associates
Manufacturer Times Facsimile Corporation
(USA) (now defunct)

The Stenafax duplicator was designed to reproduce images as well as type, using a vinyl stencil cut electronically. In designing it, Dreyfuss paid special attention to the problems of maintenance and ease of access.

The New York Times

Frigidaire Refrigerator

Raymond Loewy International Inc

Designer Raymond Loewy
Manufacturer Frigidaire Division, General
Motors Corporation, Dayton, Ohio, USA

Loewy, who was retained by the Frigidaire company from 1939, designed the Frigidaire refrigerator in 1941 and it remained in production for over 20 years. It was the best selling refrigerator in the world.

Citroën DS19
Motor Car

(in production from 1957)

Designer and manufacturer Citroën SA, Paris, France

The Citroën DS19, although it has attracted a fair share of criticism, remains one of the most celebrated pieces of modern industrial design, a masterly combination of visual flair and mechanical ingenuity. Known as the Goddess, from the phonetic accident of the pronunciation of 'DS' in French, the big Citroën carried on the reputation of the *traction-avant* Light 15 which preceded it (see page 144). Comfort was the criterion throughout its development; originally known as the 'Voiture de Grand Diffusion', this name gave the 'D' to its type number, to which was added an 'S' for 'special' and an abbreviation of the swept volume of the engine, '19' for 1911cc. It was first displayed at the Turin Motor Show, without wheels so that its startling shape could the better be admired, and it went on sale at the Paris Motor Show two years later. In 1955 its full-width body, seating within the wheelbase, separate plastics roof, large glass area, low-level air intake (without decorative grille) and almost total lack of ornament were all exceptional qualities.

National Motor Museum

Chair and Ottoman

Designer Charles Eames
Manufacturer Herman Miller Inc, Zeeland,
Michigan, USA

Office of Charles and Ray Eames

Charles Eames's famous chair and ottoman in rosewood, leather and steel for Herman Miller were designed in 1956 and have been continuously produced since then. The chair employs an ingenious swivel-tilt system and a complex method of upholstery. The covers of the cushions are zipped to a vulcanised fibre back. Inside, there is a foam rubber envelope, holding feathers and down, which snaps onto the wooden shell.

Pam Model 710 Transistor Radio

Manufacturer Pam (Radio & Television) Ltd, London, England (a subsidiary of Pye)

The Pam 710 was the first all-transistor set designed and made in Britain. The development of the transistor, which replaced the cumbersome valve, *forced* the idea of portability on radio designers. Its light weight and compactness made bulky radios a thing of the past. This early portable model shows how designers first came to terms with the challenge.

Design Council

REM Vacuum Cleaner

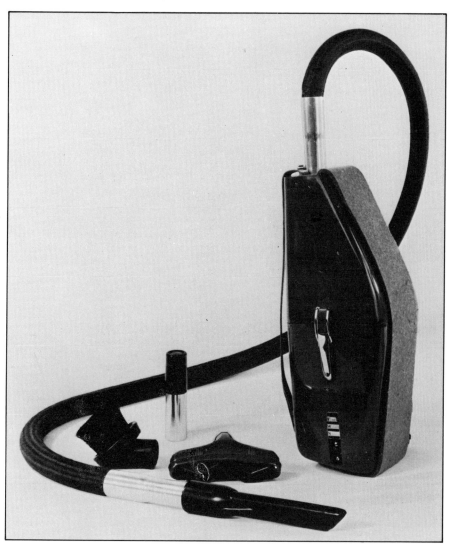

'Fotogramma', Milan

Designers Achille and Pier Giacomo Castiglioni
Manufacturer REM di Rossetti Enrico, Italy

The Castiglioni brothers had begun to experiment
with the possibilities of plastics before the Second
World War and by the mid–1950s they had
developed a considerable expertise. The pure, clean
form and the bright colours of this vacuum cleaner
are typical of the best Italian design of the period.
This design won one of La Rinascente's Compasso
d'Oro awards in 1957.

SNCASE Alouette Helicopter

Designer Raymond Loewy
Manufacturer SNCASE (Société des
Constructions Aéronautiques du Sud-Est),
Paris, France

Raymond Loewy's superb clothing for this
Alouette, Europe's most popular helicopter, was
another example of his putting an appropriate and
elegant form on what, had it been left purely to the
engineers, would have been an unrelated mass of
bits and pieces. The designer is standing on the
right.

Caper, Paris

Imperial Portable Typewriter

Designer and manufacturer Imperial Typewriter Company, Leicester, England

The 1957 model number 5 was a late development of the Good Companion series, which was introduced in 1932, and shows the influence of Italian design in Britain during the 1950s.

Science Museum

Braun Model KM321 Kitchen Machine

Designer Dieter Rams
Manufacturer Braun AG, Frankfurt, Germany

More than any other European manufacturer, except perhaps Citroën, Braun has established itself as a leader in good design. This was a conscious part of the policy that emerged from the reorganisation of the company which took place at the hands of Max Braun's sons, Artur and Erwin. Braun design is very German and very 1950s. Many of the designers employed by the company, especially Hans Gugelot and Dieter Rams, now director of their design department, were products of the austere discipline which obtained at the Hochschule für Gestaltung at Ulm. These designers produced an awesomely functional design theory – all Braun products are designed on relentless grids, and this kitchen machine is typical of them. Although a permanent display of Braun equipment was created at the Museum of Modern Art in New York in 1958, Braun design has always been too purist for popular American taste. Ironically, the company is now owned by the Gillette Company.

Commenting on the Braun machine, and disparaging the Kenwood Chef (inset), *Design* magazine said in 1959: 'The great triumph of the Braun is the sense of unity between all the components and all the attachments, marking them as the work of a single hand or a team of hands.'

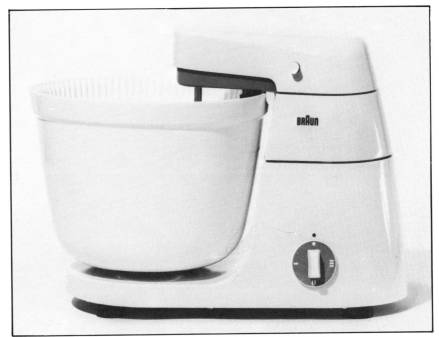

Braun AG

Kenwood Chef food mixer, c1959

Design Council

Wall Clock

Die Neue Sammlung, Munich

Designer Max Bill
Manufacturer Gebrüder Junghans, Schramberg, Germany

Max Bill was the architect of the Hochschule für Gestaltung at Ulm, the institute whose policies most closely resembled those of the Bauhaus, and a designer whose work represents the school's aesthetic policy. The Hochschule für Gestaltung insists on visual cleanness, purism and asceticism in its designs. Max Bill's exquisitely pure clock achieves all these ends.

IBM Executive Electric Typewriter

Designer Eliot Noyes
Manufacturer IBM, Endicott, New York, USA

Along with Olivetti, Braun and Citroën, IBM now has a readily identifiable model line. It is the only American corporation which can claim this distinction. That this is so is no accident. The story goes that Thomas J Watson, the IBM President, was driving through New York with his friend, the architect Eliot Noyes, when he saw the famous Olivetti showroom on Park Avenue. Watson wanted a corporate design policy so that his products, like Olivetti's, would be readily identifiable. He insisted, also, that the design should reflect the quality of the products, and he chose Eliot Noyes as a consultant to oversee all IBM design. Noyes immediately hired the designer Charles Eames and the graphic designer Paul Rand. The typewriters he kept for himself. The fore-runner of this machine appeared in 1941, although IBM had been making electric typewriters since 1933, and boasted features such as proportional spacing, which allowed a skilled typist to justify the right-hand margin. The pleasing organic shape which Noyes developed for the machine has been progressively refined and is one of the recent successes of American design.

Science Museum

Fiat 500 Motor Car

Designer Dante Giacosa
Manufacturer FIAT SpA (Fabbrica Italiana
Automobile Torino), Turin, Italy

The Fiat 500, which appeared after the *seicento* model, was one of the first small cars to incorporate the ideal of the small car package of the 1960s. Its heritage goes back to the pre-war Fiat Topolino (inset). At a time when Italian design was temporarily depressed, the baby Fiat won a Compasso d'Oro award, given by the La Rinascente store.

Design Council

National Motor Museum

Fiat 500 Smith Siata Special (Topolino) motor car, 1938

Olivetti Tekne 3 Typewriter

Designer Ettore Sottsass Jr
Manufacturer Ing C Olivetti & C SpA, Ivrea, Italy

The beautiful clean geometric forms of Sottsass'
Tekne 3 evolved into Olivetti's Praxis 48 in 1962-
63. His design is eloquent of the changing visual
criteria which were to dominate the 1960s. Perhaps
under the influence of IBM, whose high-
technology wares had recently benefited from the
new design awareness of the parent company, there
was a trend in the 1960s towards machines which
were not necessarily technologically advanced
assuming the appearance of sophisticated systems
like large computers. Olivetti, the first large manufacturer to turn typewriters into sculpture,
reacted to this immediately. The Tekne 3 is such a
machine.

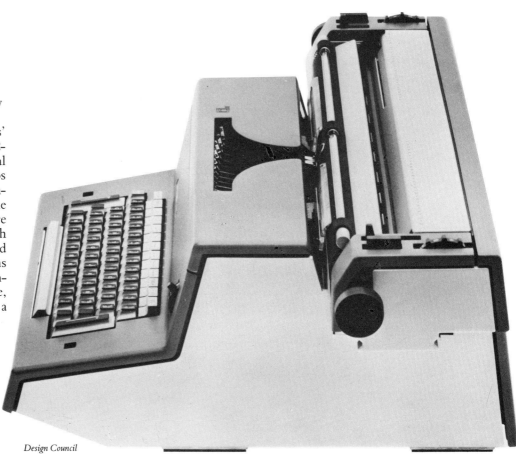

Design Council

Television Set (Project)

Design Council

Designers Hans Gugelot and Helmut Müller-Kühn

This unexecuted project by two designers working for Braun bears all the signs of being a product of the Ulm Hochschule für Gestaltung, where Gugelot worked. The muted two-tone plastics case and beautiful purity are the identifying characteristics. It was intended for the Telefunken company, but never went into production.

Braun Portable Radio-Record Player

Designer Dieter Rams
Manufacturer Braun AG, Frankfurt, Germany

This portable radio-record player, designed by Dieter Rams in the period when he was moving away from the influence of Hans Gugelot, shows Braun's geometric grid design at its severe and awesome best. One of the interesting technical features is that records are played from beneath, the stylus being engaged by moving the sliding knob at the top of the machine.

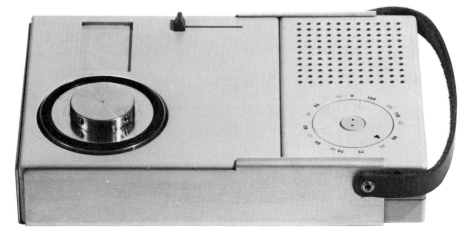

Braun AG

1959

Austin A40 Motor Car

National Motor Museum

Designer Pinin Farina
Manufacturer Austin Motor Company Ltd, Birmingham, England

It was Leonard Lord of BMC who inspired both Farina's Italian influence on BMC styling and Alec Issigonis's radically new engineering. *Design* magazine said that the A40, mechanically similar to the bulbous, dated A35, was 'the best looking product of the BMC'. Farina had established British contacts when he had become an RDI in 1957. Although the Countryman version of the A40 predicted the three-door small car cult of the 1970s, BMC failed in its engineering, its interiors and its attention to detail to match the promise of Farina's rakish styling job.

Morris Mini Minor Motor Car

Designer Alec Issigonis
Manufacturer Morris Motors, Oxford, England

When it was introduced in 1959, the Mini Minor (which was also produced as the Austin Super Seven) was the most refined product of the ideas that Issigonis had been developing with the Morris Minor. In its appearance it borrowed something from another BMC product, the Austin A35, but the novelty of the Mini was in its engineering. Features such as front-wheel drive, transverse engine, 10in wheels and rubber suspension enabled a car to be produced which combined sophisticated engineering with compact size at a modest price. It set a European standard for designing small cars which has not yet been superseded.

British Leyland

Preliminary sketch of Mini by Alec Issigonis

British Leyland

1960-61

AEG Food Mixer

Designer Peter Sieber
Manufacturer AEG, Frankfurt, Germany

The AEG food mixer – every bit as untidy as a contemporary Kenwood Chef – shows that the Braun KM321 was not austerely beautiful just because it was German, but because it was a product of a real design philosophy (see page 196). Although both the AEG and the Braun machines use similar materials, the AEG design is far less unified in conception.

AEG

1961

Braun Electric Desk Fan

Designer and manufacturer Braun AG, Frankfurt, Germany

Braun design was an expression of the 1950s. All the mechanical parts are always tastefully hidden behind a chaste housing, in this case of plastics and metal, and the form of the design is intended to be emblematic of the way the machine functions. Together with the fan, the shape of the barrel is intended to suggest the circular movement of the fan blades.

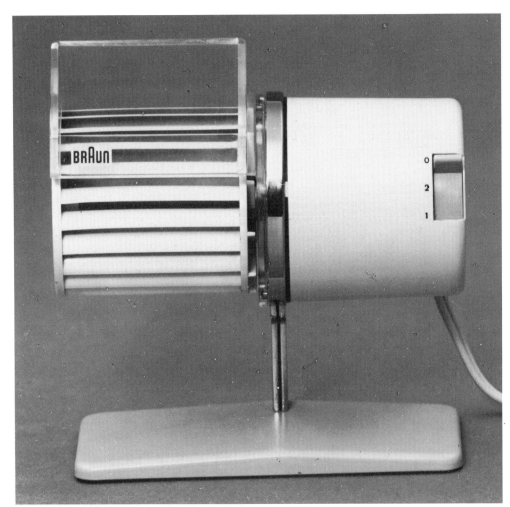

Braun AG

IBM 72 Golfball Electric Typewriter

Designer Eliot Noyes
Manufacturer IBM, Endicott, New York, USA

The replacement of type-bars by an interchangeable 'golfball' made little difference to the appearance of the office model typewriter, even though the technology was entirely new. The invention of the 'golfball' meant that all 88 characters were contained on one more or less spherical head. The only visible change this produced in the styling was that the carriage no longer moved, but the head itself travelled along the page as typing proceeded.

Design Council

1962

Braun Stereo Record Player and Radio

Designer Dieter Rams
Manufacturer Braun AG, Frankfurt, Germany

This is the Braun Audio 2 set, the latest of the company's combinations of record player and radio in a single unit. The design predicts many of the visual whims of the 1960s, including the mechanical frankness which was later taken up for popular markets by the Japanese. It is characteristic of Braun's geometric grid design, a style idiom which became known as *Schneewittchens Sarg* (Snow White's coffin) among their employees.

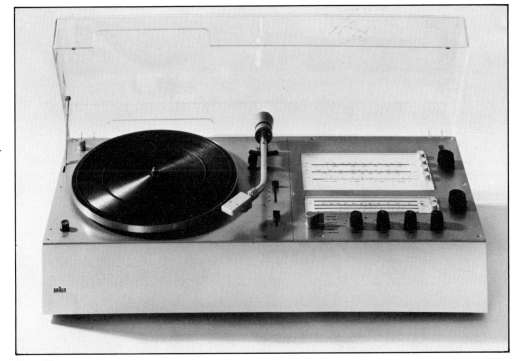

Braun AG

Siemens Norge A/S

Siemens Model 14 Stereo Tape Recorder

Designer Egil Rein
Manufacturer Siemens Norge A/S, Norway

Egil Rein's Siemens tape recorder is typical of the austere, clean style which Scandinavian manufacturers adopted for their products. Apart from the dated knurled knobs, symmetrically disposed behind a bank of three push buttons, Rein's design does not look exceptionally stale, even 16 years later. His careful use of simple materials (teak veneer, plastics and steel) and his chaste grey colour scheme contrast dramatically with contemporary British design of consumer electronics which relied largely on the iconography of the juke-box for its effects.

1962

Teknitron Siera SN59T 348A Television

Designer Jørgen Skogheim
Manufacturer Teknitron A/S, Oslo, Norway

Teak became one of the major popular styling materials of the 1960s, following the example of the Scandinavians, who employed it widely. Skogheim's television set of 1962 uses teak in its cabinet construction. It was also about this time that the screen became the dominating visual feature of the set's design and Skogheim's Siera is an early example of this mode which Bang & Olufsen, another Scandinavian manufacturer, gave European currency.

Norge/Norwegian Industrial Design by Alf Bøe, Kunstindustrimuseet, Oslo, 1963

1963

Kodak Instamatic 100 Camera

Eastman Kodak Company

Designer and manufacturer Eastman Kodak, Rochester, New York, USA

The Instamatic is the Box Brownie of the second half of the century (see page 101). Technically, nothing could be simpler. The film was loaded by cassette, avoiding the fiddly threading operation, and there was a single speed shutter and a 'Magicube' flash system. The basic geometrical shape of the plastic housing reflects the simplicity of the machine itself. For Kodak, the introduction of the more expensive cassette-loading film meant an enormous leap in turnover; the company found the public's reaction to the almost fault-free and failure-proof system was, simply, to buy more film!

Turntable

Designer David Gammon
Manufacturer Transcriptors Ltd, London,
England

David Gammon's aluminium and acrylic record
turntable has never been in series production, but it
is instructive to view it as an exquisite example of
mechanical frankness. After the plastics stylising of
the 1950s, the reaction towards a pure style dependent on a reverence for the details of the machine
was to be expected. The exposure of the mechanical parts was an entirely wilful act; ironically,
Gammon's turntable was extremely expensive. By
1964, industry was so used to disguising its products, it became more costly to expose them.

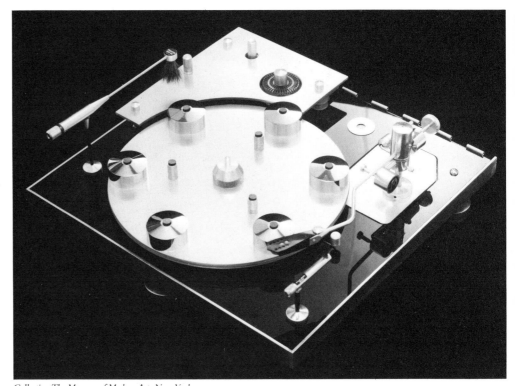

Collection The Museum of Modern Art, New York

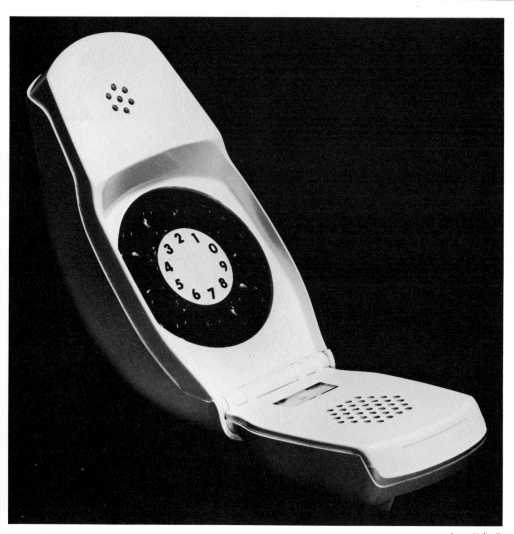

Janez Kalisnik

Plastics Folding Telephone

Designers Marco Zanuso and Richard Sapper
Manufacturer Società Italiana Telecomunicazioni
Siemens SpA, Milan, Italy

Like Joe Colombo's folding kitchens, Zanuso's and Sapper's folding telephone puts an emphasis on novelty. There is really very little advantage in having an articulated telephone, but this model is featured here, beyond the chronological sequence, because it is highly characteristic of some assumptions made about design after the purity which emerged in the best products of the 1950s.

BIOGRAPHIES

The study of design history, at least at the documentary level, is in a primitive state. The creators of some of the most astonishing designs made this century are largely unknown to us because the companies who employed or retained them have, under commercial pressure, often been forced to destroy the space-consuming archive material which describes their careers. This is as true of a corporation the size of General Motors as it is of a modest manufacturer of bicycles or steel chairs.

This biographical list is deliberately a brief one, but it could hardly have been otherwise. The material for compiling a comprehensive list of twentieth-century industrial designers is hardly yet available. For the time being, this list is restricted to those designers about whom it is actually possible to say *something*. It would have been otiose to include here individuals, perhaps mentioned somewhere in captions to the photographs, about whom it is often impossible to say more than that he or she, for instance, 'studied at the Bauhaus'. Furthermore, in the case of architects mentioned here, the entries make no attempt to rival or to duplicate information already available about them in standard reference works.

Alvar Aalto

Peter Behrens

Architect and furniture designer
b Kuortane, Finland, 1898 d Helsinki, Finland,
1976

Of all leading architects associated with the
Modern Movement, Alvar Aalto has, perhaps,
written the least. His theories and his attitude to
his work are relatively unknown, although his
most famous buildings, the Sanatorium at
Paimio (1929-33) and the Library at Viipuri
(1935) are among the most familiar images of
modern architectural publishing. Aalto was also
the creator of some strikingly simple and practical plywood furniture for the Huonekalu
Company which grew out of his close association with the special characteristics of his chosen material.

Bibliography
Karl Flieg *Alvar Aalto* Thames & Hudson, 1975.
Reyner Banham 'The One and the Few: the Rise
of Modern Architecture in Finland', *Architectural
Review* April 1957, pp 243-259.
Alvar Aalto 1898-1976 exhibition catalogue, The
Museum of Finnish Architecture, 1978.

Photo Artek

Architect and industrial designer
b Hamburg, Germany, 1868 d Berlin,
Germany, 1940

Peter Behrens is the outstanding industrial
designer of the early years of the twentieth century. He was trained at the Karlsruhe and Düsseldorf Art Schools and joined the Munich Sec-
ession in 1893. From 1907 he was in charge of
factory building, workers' housing and industrial design for the enormous Allgemeine Elek-
tricitäts-Gesellschaft in Berlin where, as one of
the first architect-designers to be employed by a
large industrial concern, he established a design
identity for the company, doing jobs as diverse
as the famous Turbine assembly hall in Berlin
(1909), the AEG logo and electrical appliances.
Although Behrens' work for AEG was pure and
austere enough for him to merit inclusion
among Pevsner's *Pioneers*, when allowed to
work on a more traditional architectural project,
such as the German Embassy at St Petersburg
(1911; later demolished), he turned out a
run-of-the-mill neo-classical design.

At one time or another Behrens had as pupils
Walter Gropius and Mies van der Rohe and,
very briefly, Le Corbusier. He was Director of
Architecture at the Vienna Academy in 1932 and
was Director of the Department of Architecture
at the Prussian Academy in Berlin from 1936
until his death.

Bibliography
F Hoeber *Peter Behrens* Munich, 1913.
Julius Posener 'L'Oeuvre de Peter Behrens',
L'Architecture d'Aujourd'hui March 1934, pp 8-
29.

P Morton Shand 'Peter Behrens', *Architectural Review* September 1934, pp 39-42.
Ernesto Rogers et al 'Numero dedicato a Peter Behrens', *Casabella Continuità* Special number, no 240, 1960.
Industrie Kultur – Peter Behrens und Die AEG 1907-1914 exhibition catalogue, IDZ, Berlin, 1978.

Photo AEG

Norman Bel Geddes

Industrial designer
b Adrian, Michigan, USA, 1893 d New York, USA, 1958

Norman Bel Geddes was born of Scottish and German ancestry, was irregularly schooled and ultimately expelled. He was briefly a student at the Art Institute of Chicago but, on meeting the Norwegian painter Hendrik Lund, he decided to become independent and went his own way. First of all he worked as an advertising draughtsman but, like Henry Dreyfuss, he developed an interest in the theatre. In 1927 he took up industrial design and his first project was to design car bodies for the Graham-Paige Company which were, according to his brief, to look five years ahead of their time. Although Bel Geddes's designs were aesthetically pleasing and predicted many later styling idioms (such as faired-in mudguards), they did not please the Graham-Paige directors and were never used. By 1934, however, Bel Geddes's career as an industrial designer had taken off.

The theatre always played an important part in Bel Geddes's life and career. He was an idealist and a visionary, two characteristics which his unexecuted projects betray. However, he also had a supremely practical turn of mind, stressing that design was just a matter of thinking and that drawings – no matter how persuasive they might be to the client – should always come last in the design process. His procedure was first of all fully to understand any given problem and then to experiment patiently and diligently until the best solution presented itself. Although Bel Geddes worked on such famous products as the Chrysler Airflow – where he was called in to do a new job on the nose to boost declining sales – and some Electrolux appliances, his main contribution to industrial design in this century has been in the realms of visual ideas.

Bibliography
Norman Bel Geddes *Horizons* Little, Brown, Boston, 1932 (English edition, John Lane, The Bodley Head, 1934), reprinted Dover, New York, 1978.
Norman Bel Geddes *Magic Motorways* Random House, New York, 1940.
Kenneth Reid 'Masters of Design 2: Norman Bel Geddes', *Pencil Points* vol 18, January 1937, pp 2-32.
Norman Bel Geddes (ed William Kelley) *Miracle in the Evening* Doubleday, New York, 1960.
Arthur J Pulos 'The Restless Genius of Norman Bel Geddes', *Architectural Forum* July-August 1970, pp 46-51.

Photo University of Texas and Edith Lutyens Bel Geddes, executrix of the Norman Bel Geddes Estate

Mario Bellini

Harry Bertoia

Industrial designer
b 1935

Mario Bellini, working largely for Olivetti, is among the most prominent of the younger generation of Italian designers. His work is characterised by the use of smooth curves on shapes which are fundamentally rectangular.

Bibliography

Edilizia Moderna no 85, 1964 pp 9-11.
Paolo Fossati *Il Design in Italia 1945-1972* Einaudi, Turin, 1972.
James Woudhuysen 'The Typewriter as Just Another Limb', *Design* no 348, December 1977, pp 50-51.

Photo Mario Mulas

Industrial and furniture designer
b San Lorenzo, Italy, 1915 d Barto, Pennsylvania, USA, 1978

Harry Bertoia arrived in the United States from Italy in 1930 and studied art in Detroit. He worked variously for Evans Product Co, the US Electronics Laboratory and as a jewellery designer. He is best known for his furniture designs made for Knoll International, where he was a chair designer from 1950 to 1954 and produced his popular steel wire lattice chair. He also worked as a sculptor and his work is displayed at the General Motors Technical Centre at Warren, Michigan and at Dulles Airport, Virginia.

Bibliography

Clement Meadmore *The Modern Chair – Classics in Production* Studio Vista, 1974.
Karl Mang *Geschichte des Modernen Möbels* Gerd Hatje Verlag, Stuttgart, 1978.

Photo Form International

Pierre Boulanger

Automotive engineer
b 1886 d 1950

Although almost unknown himself, Pierre Boulanger's work for Citroën cars is known across the world. While associated with Michelin, Citroën's parent company, Boulanger took over the car manufacturer's engineering division. That was in 1935. Immediately he began the design programme that led to the creation of the timeless 2CV. Boulanger was killed in a road accident in 1950 while engaged on development work for the new big Citroën, the DS19, which was to appear in 1955.

Photo Citroën SA

Marianne Brandt

Metalworker
b 1893

Marianne Brand, with Wilhelm Wagenfeld, was perhaps the best known product of the Bauhaus' Metal Workshop, which was run by László Moholy-Nagy. Although she was in no sense an industrial designer, Marianne Brandt's designs for teapots and other domestic metalware became very well known and were influential in developing a taste for modern design across all of Europe.

Bibliography

Hans Maria Wingler *The Bauhaus* MIT Press, Cambridge, Mass, 1968.

Photo Galerie am Sachsenplatz, Leipzig/DDR

Marcel Breuer

Architect and furniture designer
b Pécs, Hungary, 1902

Marcel Breuer, like Moholy-Nagy before him, took the traditional route out of backward Hungary and arrived in Vienna in 1920, intending to become a painter. He soon moved to Weimar, where he became a student at Walter Gropius' Bauhaus. By 1924 he was in charge of the furniture workshops there and by 1928 he had set up his own studio in Berlin. In 1933, with the brothers Alfred and Emil Roth, he designed a block of flats in the Doldertal district of Zürich for the art historian Siegfried Giedion, which became one of the best known examples of the International Style in architecture. From 1936 to 1937 Breuer shared a practice in England with the architect and writer F R S Yorke, and during this time he met the furniture manufacturer Jack Pritchard. Breuer's most famous examples of design have been the tubular steel chairs he designed at the Bauhaus, and the elegant plywood chaise longue he made for Pritchard's Isokon concern. Breuer now lives in New York and has latterly been associated with IBM under the design programme started for them by Eliot Noyes. He has designed laboratories for IBM at La Gaude, France and Boca Raton, Florida.

Bibliography

Marcel Breuer *Marcel Breuer 1921-1962* Gerd Hatje Verlag, Stuttgart, 1962.
Hans Maria Wingler *The Bauhaus* MIT Press, Cambridge, Mass, 1968.

Photo Bauhaus-Archiv

Achille Castiglioni

Industrial designer
b Milan, Italy, 1918

Achille Castiglioni, with his brother Pier Giacomo, has been among the leading Italian designers since immediately before the Second World War. The two have specialised in the field of lighting, although they have also designed other electrical appliances, and have exhibited at every Milan Triennale since 1947.

Bibliography
Edilizia Moderna no 85, 1964, pp 12-14.

Photo Stile Industria *magazine*

Serge Chermayeff

Architect and industrial designer
b Caucasus, Russia, 1900

Serge Chermayeff was educated in Britain. He went to Harrow School and moved directly into journalism, changing only later to architecture, industrial and furniture design. While Erich Mendelsohn was in Britain, Chermayeff was his partner and together they created the De La Warr Pavilion at Bexhill in Sussex (1935), one of the earlier examples of the International Style in British architecture. At about the same time Chermayeff established contacts with the E K Cole company, for which he designed radio cabinets which were aesthetically well ahead of their time. He also designed studio interiors for the new BBC building, Broadcasting House, in Langham Place, London. Chermayeff now lives in Massachusetts.

Bibliography
Serge Chermayeff and Christopher Alexander *Community and Privacy: Towards a New Architecture of Humanism* Pelican, Harmondsworth, 1966.

Photo Condé Nast

Wells Coates

Joe (Cesare) Colombo

Architect and industrial designer,
b Tokyo, Japan, 1895 d Vancouver, Canada, 1958

Wells Coates's Canadian mother had been a pupil of Louis Sullivan. The combined influences of the 'functionalist' stream coming through her, and the experience of an Oriental childhood could be seen to have affected Coates's view of the world and the style of the buildings and the appliances which he designed. Coates was a polymath. He gained a PhD from London University in 1924 for work on diesel engines, and promptly became a journalist on the *Daily Express*. He first came into contact with contemporary modern architecture during visits to Paris and became a founder member of MARS (Modern Architecture Research Group), the English branch of CIAM (Congrès Internationaux d'Architecture Moderne) in 1933. Besides being the first architect to introduce the International Style of architecture to England, which he did with his Lawn Road flats, built for Jack Pritchard in Hampstead in 1934, Coates, like Chermayeff, worked on radio cabinet design for the E K Cole company both before and after the Second World War.

Bibliography

Sherban Cantacuzino *Wells Coates* Gordon Fraser, 1978.

Photo Barratt's Photo Press Ltd

Architect and furniture designer
b 1930 d 1971

Joe Colombo, one of the darlings of the smart Italian magazines, was an architect who specialised in producing space-saving, compact, ingenious designs for stowaway kitchens, wardrobes and studies, although perhaps his most famous piece of design is his curved Perspex lamp, with lens and reflector combined, which he produced for O-Luce of Milan. Colombo's work has not been vastly influential, but is typical of a sort of self-conscious modishness characteristic of much of current Italian design.

Bibliography

'Una Nuova Concezione dell'Arredamento: Joe Cesare Colombo', *Lotus* vol 3, 1966-67, pp 176-195.

Photo Desmond O'Neill

Henry Dreyfuss

Industrial designer
b Brooklyn, New York, USA, 1904 d Pasadena, California, USA, 1972

During the early 1920s Henry Dreyfuss, who was to become one of the big three American industrial designers, worked on Broadway designing stage sets. His first – unsuccessful – design consultancy was with a department store and he opened an office on his own account in 1929, while still working on Broadway sets. Among the blue-chip clients Dreyfuss worked for were Bell Telephones and Lockheed Aircraft. For the latter he designed the interior of the Super Constellation airliner.

In all Dreyfuss's work there is a persistent interest in people and in relating all design projects to the human scale. This is apparent in two separate areas of the firm's concern: even today Henry Dreyfuss Associates restricts its business to no more than 15 clients; and, together with this concern to keep the firm's relations at an almost personal level, many of the design problems which consume their attention are those related to the human figure and how it works best. Indeed, of all the great industrial designers, Dreyfuss could be said to have pioneered the science of anthropometry. His own book, *Designing for People*, stressed the importance of human engineering and this interest in human performance continues in his office today: Niels Diffrient, one of the partners in the firm, has recently published a book called *Humanscale*, an attempt to make scientific all the designer's observations about human performance.

Bibliography

Henry Dreyfuss *Designing for People* Simon & Schuster, New York, 1955; second expanded edition published as *The Measure of Man*, 1960.

Photo Design Council

Charles Eames

Architect and furniture designer
b St Louis, Missouri, USA, 1907 d St Louis, Missouri, USA, 1978

Charles Eames enjoys almost a cult reputation among architects and designers, although the significance of his rather small output is, perhaps, less tangible than the dimensions of his critical reputation. His best known designs are the chairs made for Herman Miller, especially the plywood and leather swivel-joint chair and ottoman, introduced in 1956. Originally Eames practised architecture privately in the mid-West and became a freelance designer of furniture, toys, films and buildings. He was among the architects and designers selected by Eliot Noyes to contribute to IBM's new corporate design programme. He was a regular contributor to the annual IDCA conferences in Colorado and is well known for a number of oracular, gem-like sayings, such as his claim that the London taxi cab is the best piece of industrial design in the world.

Bibliography

Whole issue of *Architectural Design* devoted to Eames's work, vol 36, September 1966.

Photo Office of Charles and Ray Eames

Harley J Earl

Automotive engineer
b Los Angeles, California, USA, 1893 d 1954

Harley J Earl was vice-president of General Motors in charge of the styling staff during the period when manufacturers of automobiles became conscious of the importance of styling. In terms of sheer aggregates of numbers, it is probably true to say that Earl has had as much influence on the way the world looks as any man who has ever lived.

He was born in California, the son of a coachbuilder, and educated at Stanford University. From 1927, when he became director of the art and colour section, to 1937, when he became director of styling, and then throughout the 1940s and early 1950s, Earl dominated the aesthetic policies of General Motors, the largest industrial undertaking in the world. In this position, his personal skills and tastes affected the appearance not only of cars and trucks, but also of machines (fans, radios and refrigerators, for instance) produced by the corporation.

Earl's contribution to the popular understanding of what is 'modern' can scarcely be calculated. In the 1950s he was happy to claim personal responsibility for the styling of 31 million cars. Under him General Motors not only developed the annual styling inflection (for instance, the meretricious wrap-around windscreen), but also refined the designed obsolescence policy which was its concomitant. It was also under Earl's direction that General Motors began exhibiting 'dream' cars at public fairs to test out new ideas on the popular taste. Although his extravagant designs for dream cars like the Buick Le Sabre and the Pontiac Strato-Streak (the latter being the model for the British Vauxhall Cresta of 1958 to 1961) – modelled at once on both science fiction and the vocabulary of racing car design – certainly caught the mood of the time, Harley Earl, despite his influence, has never received general recognition as a creator of an automotive iconography.

Photo General Motors

Fritz Eichler

Industrial designer
b 1911

Fritz Eichler studied art history and drama in Berlin and Munich, and worked in theatre design from 1945 to 1963. He was the genius who, with Artur and Erwin Braun, brought the old Max Braun company to the forefront of industrial concerns which enjoyed an international reputation for good design. It was Eichler who established the links between the new Braun AG and the Hochschule für Gestaltung at Ulm which were so influential on the appearance of all Braun products from the mid-1950s on.

Photo Braun AG

Dante Giacosa

Hans Gugelot

Automotive engineer
b Rome, Italy, 1905

Dante Giacosa's products are better known than the man. As director of Fiat's Progetti e Studi Autoveicoli, Giacosa was largely responsible for some of the most enduring and successful of modern European small cars, the Fiats 1100, 1400, 1900, 125, 128 and, most loved of all, the Nuova 500. Giacosa joined Fiat in 1928 and has become a member, or an honorary member, of many national engineering bodies. He is also a prolific writer and public speaker on engineering matters.

Photo Barratt's Photo Press Ltd

Industrial designer
b Makasar, Indonesia, 1920 d Frankfurt, Germany, 1965

Hans Gugelot, known familiarly as 'Gütsch' to those admirers who saw in his austere designs for Braun and others an exquisite symbol of the best in mid-twentieth-century design, was one of the most influential industrial designers of his generation. Gugelot began his work for Braun after he met Fritz Eichler in 1954, and he made the principles of the Ulm Hochschule für Gestaltung a commercial reality for the Braun company. Although he effectively created the international image of Braun products, towards the end of his life Braun's design policy – which he had done so much to define – was beginning to move away from the standards he had set out for it. As well as his work for Braun, Gugelot also designed prefabricated housing at Zürich Witikon.

Bibliography
Alison and Peter Smithson 'Concealment and Display: Meditations on Braun', *Architectural Design* vol 36, July 1966, pp 362-363.
Reyner Banham 'Household Godjets', reprinted from an original article in *New Society* in Paul Barker (ed) *The Arts in Society* Fontana, 1977, pp 164–170.

Photo Design Council

Alec Issigonis

Edgar Kaufmann Jr

Automotive engineer
b Smyrna, Turkey, 1906

Alec Issigonis is one of the few British automotive engineers to have achieved anything like celebrity. He was educated at Battersea Polytechnic and soon became a draughtsman with Rootes Motors in Coventry. He joined Morris Motors in Oxford and became its chief engineer, becoming successively chief engineer of the British Motor Corporation and its technical director in 1961. Issigonis became an RDI in 1961 and is now a consultant to British Leyland. Issigonis is famous for the design of three highly influential small cars, the Morris Minor (1948), the Morris Mini Minor/Austin Super Seven (1959) and the Morris 1100 (1962). Each of these cars was pioneering in a special way; the first two were conceived almost totally by Issigonis, but the 1100 of 1962 used bodywork styling derived from the Farina model which had been used on other BMC products, like the Austin A40. Of these three cars, the Mini is still in production and has become the most successful British car ever made. It created a world demand for sophisticated, small front-wheel-drive cars which has still not been satisfied. For once, with Issigonis's Mini design, the British motor industry led the world. The design Issigonis established for the Mini in 1959 has been imitated directly by every major motor car manufacturer, with the exception of the American Chrysler and General Motors Corporations.

Photo British Leyland

Writer

Edgar Kaufmann Jr, through a long association with the Museum of Modern Art in New York, for which he wrote a number of popular booklets, including *What is Modern Design?*, was one of the leading American modern design propagandists of the older generation. He was never particularly revolutionary or innovative, and his opinions have always represented the median in discussions about design. Lately, Kaufmann has come to question the unswerving avoidance of ornament in modern design, one of the articles of faith of the Modern Movement in design which he had done so much to promulgate.

Bibliography

Edgar Kaufmann Jr 'Borax, or the Chromium-Plated Calf', *Architectural Review* August 1948, pp 88-93.
'Design sans Peur et sans Ressources', *Architectural Forum* September 1966, pp 68-70.
'Modern Design does not need Ornament', *College Art Journal* vol 6, Winter 1946, pp 140-142.
'The Design Shift 1950-1960', *Industrial Design* vol 7, August 1960, pp 50-51.
'What is Modern Design?', *Museum of Modern Art Bulletin* vol 14, Fall 1946, pp 3-13.

Raymond Loewy

Industrial designer
b Paris, France, 1893

Raymond Fernand Loewy arrived in the United States in 1919, a penniless captain in the French army. He had amused himself on the transatlantic crossing by making sketches of his fellow passengers. These sketches so impressed Sir Henry Armstrong, the British Consul in New York, that he put Loewy in touch with Condé Nast, publishers of *Vogue* and *Harper's Bazaar* magazines. Loewy began work as a fashion illustrator and also worked with the producer and impresario Florenz Ziegfeld.

He founded his own design studio in 1927 and this was chosen by *Life* magazine as number 87 in the 100 events that had shaped America. Raymond Loewy Associates rapidly became the biggest industrial design firm in the world. With valuable contracts and retainers from many of the biggest American corporations, it was possible, during the 1940s and 1950s, to spend almost all of every day in contact with a product designed by Loewy. During the Second World War it was estimated that objects designed by Loewy were turning over more than $900,000,000 a year.

Loewy's style is inimitable, but his success has produced much criticism. His detractors claim that Loewy is merely a flamboyant stylist, although he refutes this vigorously. Loewy claims that the complete vindication of his design theories was the commission from NASA to design the interiors of the Skylab spaceship. Here, austere function had to be the most important consideration, although elsewhere in his work Loewy has been predominantly interested in the philosophical and commercial usefulness of beauty. These ideas were all expressed in his witty autobiography, *Never Leave Well Enough Alone*.

Bibliography

Raymond Loewy *Never Leave Well Enough Alone* Simon & Schuster, New York, 1951.
Raymond Loewy *La Laideur se Vend Mal* Gallimard, 'L'Air du Temps', no 8, Paris, 1963 (translation of *Never Leave Well Enough Alone* by Miriam Cendrars, with additional material).
The Designs of Raymond Loewy exhibition catalogue, Renwick Gallery, Washington, 1976.
Raymond Loewy *Industrial Design* The Overlook Press, New York, 1979.

Photo Stephen Bayley

Mies van der Rohe

Architect and furniture designer
b Aachen, Germany, 1886 d Chicago, USA, 1969

Ludwig Mies van der Rohe is well known as the most purist of all modern architects, the one most dedicated to the beauty of simplicity and high finish. He said of his work that he didn't want it to be interesting, but that he just wanted it to be good, an opinion peculiarly revealing of his passions. With oracular statements like this, and his even better known *diktat* that 'less is more', it will be seen that Mies' architectural philosophy had much in common with the tradition of German industrial design. Mies joined the Werkbund in 1925, after training with Peter Behrens from 1908 to 1912 and being a member of the 'Ring' group of radical architects. In 1927 he participated in the housing exhibition at Weissenhof near Stuttgart and in 1929 he designed the German pavilion at the Barcelona World's Fair. It was for this pavilion that he also designed the 'Barcelona' chair; although the pavilion has long since been destroyed, the chair is still in production today and is, in fact, used inside his own Neue Nationalgalerie in West Berlin. Mies was the third and last Director of the Bauhaus in its most difficult phase, under the shadow of the Nazis. Under him the school closed in Berlin in 1933 and Mies emigrated to the United States in 1938. He became Director of Architecture at the Illinois Institute of Technology in 1948. It was through Mies in Chicago and Gropius at Harvard that the ideas of the European Modern Movement in architecture and design were established across the Atlantic.

Bibliography
Arthur Drexler *Mies van der Rohe* Braziller, New York, 1960.
The Chairs of Mies van der Rohe Museum of Modern Art, New York, 1976.
Werner Blaser *After Mies* Van Nostrand Reinhold, New York, 1978.

Photo Bauhaus-Archiv

William Mitchell

Automotive engineer
b Cleveland, Ohio, USA, 1912

Bill Mitchell, the protégé of the great General Motors car stylist Harley Earl, was hired by the Corporation in 1935 and soon became head of the Cadillac studio. He aimed in his designs to capture the magic of low-volume cars like the Duesenberg and the Stutz and of high quality European models, like the Mercedes-Benz. This was a different emphasis from Earl's, which employed the idioms of the race track and the pulp magazine in its automotive iconography. Mitchell was vice-president in charge of styling for General Motors from 1958 to 1978, when he retired to run his own design consultancy.

Photo General Motors

László Moholy-Nagy

Painter, photographer, film-maker, typographer and educationist
b Borsod, Hungary, 1895 d Chicago, USA, 1946

Moholy-Nagy never designed anything that has gone into series production, but he has been a very influential educationist in design. He studied law in Budapest. After war service, when he was injured, Moholy began private sketching and published an avant-garde journal, called *Ma* (Tomorrow) with his friend Ludwig Kassak. He was in Berlin by 1920 and had joined the Bauhaus staff by 1922, where he ran the metal workshop and made abstract and semi-abstract films as well as writing, designing and editing the *Bauhausbücher*. In the context of this book, Moholy is most interesting as being among the very first 'fine' artists to draw attention to the beauty of machines and to employ industrial techniques in his art. His *Buch neuer Künstler*, for instance, used photographs of aeroplanes to explain the principles of the new aesthetic which Moholy was proposing. Moholy was briefly in London before going to Chicago where, under the patronage of Walter Paepcke (the owner of the Container Corporation of America) he founded the New Bauhaus and, after it had folded, the Institute of Design. Moholy's publications are still used in design education, and have been consistently influential.

Bibliography
László Moholy-Nagy *The New Vision, from Material to Architecture* W W Norton, New York, 1938 (English language edition of Moholy's Bauhausbuch *Von Material zu Architektur*).
László Moholy-Nagy *Vision in Motion* Paul Theobald, Chicago, 1947.
Richard Kostelanetz *Moholy-Nagy* Allen Lane, 1972.
Lucia Moholy *Marginalien zu Moholy-Nagy: Dokumentarische Ungereimtheiten* Verlag Richard Sherpe, Krefeld, 1972.

Photo Bauhaus-Archiv

George Nelson

Architect and furniture designer
b Hartford, Connecticut, USA, 1908

George Nelson, through his association with the IDCA and the Herman Miller organisation, has been one of America's most energetic native apologists for modern design. He was educated at Yale and while a Fellow of the American Academy in Rome he had the opportunity to travel round Europe, meeting face to face the architects who were creating the International Style. He became co-managing editor of *Architectural Forum* and has been president of his own company, George Nelson & Co Inc, since 1947. From 1944 to 1965 he was in charge of design for the Herman Miller organisation and, under his influence, many successful furniture designers, including Charles Eames, were brought in to work for the company. He is the author of a well respected book, *Problems of Design*.

Eliot Noyes

Bibliography
George Nelson 'Business and the Industrial Designer', *Design* vol 1, no 11, November 1949, pp 2–8.
George Nelson *Problems of Design* Whitney Library of Design, New York, 1957 (2nd edition, 1965).
George Nelson *How to See – Visual Adventures in a World God Never Made* Little, Brown, Boston, 1977.

Photo Pentagram

Architect and industrial designer
b Boston, Massachusetts, USA, 1910 d New Canaan, Connecticut, USA, 1977

Eliot Noyes was a design consultant to a number of big American corporations, including IBM, Mobil, Westinghouse and Pan-American. Noyes studied architecture at Harvard, where his father taught English. He reacted against the Beaux-Arts training he had there when, like so many other young architects, he discovered Le Corbusier. He was briefly director of the Department of Industrial Design at the Museum of Modern Art in New York, following a recommendation by Walter Gropius. After the Second World War he joined Norman Bel Geddes & Company as a designer, but soon started up his own company as a designer and architect.

It was with IBM that Noyes really made his name as a force in industrial design. He was retained by Thomas Watson, the Corporation's president, to give IBM an identifiable *style*. This was in conscious imitation of the example of Olivetti (whose products Watson deeply admired) and, less consciously, in imitation of AEG. As well as designing some equipment himself, Noyes was also the architect of IBM's laboratory and education centre at Poughkeepsie, New York, and Management Development Centre at Armonk, New York. Under Noyes's direction IBM's Real Estate and Construction Division created an international series of buildings which reflected the Corporation's enthusiasm for good design. Noyes hired Marcel Breuer to design buildings for IBM Finance at La Gaude and Boca Raton, Florida. He established a uniform product style for

Marcello Nizzoli

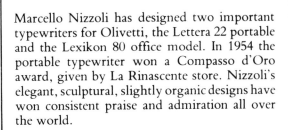

Industrial designer
b 1910

Marcello Nizzoli has designed two important typewriters for Olivetti, the Lettera 22 portable and the Lexikon 80 office model. In 1954 the portable typewriter won a Compasso d'Oro award, given by La Rinascente store. Nizzoli's elegant, sculptural, slightly organic designs have won consistent praise and admiration all over the world.

Bibliography
Paolo Fossati *Il Design in Italia 1945-1972* Einaudi, Turin, 1972.
Germano Celant *Marcello Nizzoli* Edizioni Comunita, Milan, 1968.

Photo British Olivetti Ltd

IBM's products, a necessary step for a company which relied so heavily on international complementation of its manufactures. Noyes designed for IBM's Office Products Division, and hired Charles Eames to design Special Products. He chose Paul Rand to design the corporation's graphics.

Noyes's achievement was to give a bewilderingly complex product range an identifiable style where visual anarchy had prevailed before. Although he designed relatively little himself, his was the controlling mind behind the visual rebirth of IBM.

Bibliography

Architectural Review May 1957, p 362.

Photo Derry Noyes Craig

Dieter Rams

Industrial designer
b Wiesbaden, Germany, 1932

It was Dieter Rams who, with Hans Gugelot, gave form to the revival of the Braun company in the 1950s. Gugelot brought the discipline of the Ulm Hochschule für Gestaltung into industry, while Rams developed an 'in house' style. Together they created the modern image of the Braun company. Rams' austerely beautiful product designs gave the Frankfurt electrical manufacturer an imagery to carry it through the 1960s. What was remarkable was that Rams invented this imagery a decade before: the Braun kitchen machine Rams designed in 1957 was still in production 20 years later and continues to be a commercial success in changing world markets.

Dieter Rams was apprenticed as a joiner and then studied architecture and design at the Wiesbaden Werk-Kunstschule. His first employment was with the architectural firm of Otto Apel and at the same time he joined the Deutscher Werkbund, also becoming a member of the Rat für Formgebung and an Honorary RDI.

In 1955 Rams was appointed chief designer with Braun AG and under his direction almost the whole of the Braun product range was so transformed that by 1959 a large part of it was on display at the Museum of Modern Art in New York. Rams' style and his ideology – he has said 'I regard it as one of the most important and most responsible tasks of a designer today to help clear the chaos we are living in' – have been very much influenced by the most advanced design of the period before the Second World War: in his work can be seen exactly how the Modern Movement hit industry.

Bibliography

Dieter Rams 'And That's how Simple it is to be a Good Designer', *Designer* September 1978, pp 12-13.

Photo Braun AG/A Tüllman

Egil Rein

Mechanical engineer and industrial designer
b Oslo, Norway, 1913

Egil Arne Rein is one of the few Scandinavian designers actually known by name. He has worked as construction manager for the Norwegian wing of the electrical conglomerate Siemens Norge A/S and, as a freelance industrial designer, he has made a major contribution to the establishment of a Scandinavian style in consumer products such as tape recorders and dictating machines.

Bibliography
Alf Bøe *Norsk/Norwegian Industrial Design* Kunstindustrimuseet, Oslo, 1963.

Photo Egil Rein

Gordon Russell

Furniture designer and administrator
b London, England, 1892

Gordon Russell's furniture manufacturing business at Broadway in Worcestershire has continued the highest standards of workmanship with a no-nonsense approach to design and form which owes something to the theories of the Modern Movement in design, at least as far as these were understood in England. Russell is included here because of the influence he has had on the establishment of design as a dignified, legitimate pursuit in this country. As a member of the Utility Furniture Advisory Committee, and as an original committee member both of the Festival of Britain and of the Council of Industrial Design (he was its Director from 1947 to 1959) he has made the business of good design a matter of popular concern.

Bibliography
Sir Gordon Russell *A Designer's Trade* Allen & Unwin, 1968.

Photo Design Council

Peter Sieber

Industrial designer
b 1911 d 1965

Peter Sieber had the difficult job of following in the footsteps of Peter Behrens, as he was only the second industrial designer to be employed by AEG when design resumed some significance for the company in the 1950s. Sieber founded AEG's Institut für Produktgestaltung and wrote, occasionally, for the English *Design* magazine.

Bibliography
Peter Sieber 'Record Packaging', *Design* no 77, May 1955, pp 38-42.
Peter Sieber 'Waste on the Rocks', *Design* no 80, August 1955, pp 32-34.

Photo AEG

Jørgen Skogheim

Industrial designer
b 1926

Jørgen Skogheim, a qualified cabinet-maker, is one of the Scandinavian designers who has established a contemporary language of television design, particularly in his work for the electrical company Norsk A/S Philips.

Bibliography
Alf Bøe *Norsk/Norwegian Industrial Design* Kunstindustrimuseet, Oslo, 1963.

Ettore Sottsass Jr

Architect and industrial designer
b Innsbruck, Austria, 1917

Sottsass' father studied architecture in Vienna under Otto Wagner, while Sottsass himself studied at Turin Polytechnic. He has become one of the leading Italian designers in the post-war period, the creator of clothes for machines which have all been profoundly original and have done much to condition acceptance of novel types of equipment and machinery in the home and at work. During the 1950s and 1960s Sottsass worked principally as a consultant to the electronics division of Olivetti, a company for whose paternalistic philosophy he has much sympathy. Sottsass believes that machines should be used to liberate man, not to condition him. As well as having designed typewriters such as the Tekne and the Praxis, Sottsass has also designed complete office systems for Olivetti.

Bibliography
Edilizia Moderna no 85, 1964, p 20.
Ettore Sottsass Jr – de l'Objet Fini à la Fin de l'Objet exhibition catalogue compiled by François Burkhardt et al, Centre de Création Industrielle, Paris, 1976.
Pier Carlo Santini 'Introduzione ad Ettore Sottsass', *Zodiac* no 11, 1963.
Paolo Fossati *Il Design in Italia 1945-1972* Einaudi, Turin, 1972.

Photo Ing C Olivetti & C SpA

Mart Stam

Walter Dorwin Teague

Architect
b Purmerend, Holland, 1899

Mart Stam, like his fellow countryman J J P Oud, was one of the most uncompromising devotees of the new architecture which emerged just after the First World War. With Emil Roth and Hans Schmidt he ran the magazine *ABC* from Zürich, a journal committed to the solution of social problems by the use of technology and modern architecture. According to Heinz Rasch, it was Stam who was the first person to design a tubular steel chair, in 1926, before both Breuer and Mies produced their designs. Stam was in Russia from 1930 to 1934. From 1939 to 1948 he was director of the Institute voor Kunstnijverheidsondernijs (Institute of Applied Arts) in Amsterdam. He retired from active life in 1966. Although a talented and persuasive architect and designer, Stam's inflexible politics and obdurate 'functionalism' prevented him from achieving real fame or influence.

Bibliography

Mart Stam – Documentising his Work 1920-1965, RIBA Publications, 1970.

Photo The Bauhaus *by Hans Maria Wingler, MIT Press, Cambridge, Mass, 1968*

Industrial designer
b Decatur, Indiana, USA, 1883 d 1960

With Loewy and Norman Bel Geddes, Walter Dorwin Teague established the industrial designer as an American folk hero. Teague studied at the Art Students League in New York and then worked for an advertising agency, employed first of all on grand pianos. In July 1926 he abandoned advertising draughtsmanship and set up his own studio. It was his experience with commerce that gave Teague's career its unique flavour, produced out of a happy marriage of business acumen with high artistic ability. His first important job came in 1927 from Eastman Kodak and gave him the opportunity to put into practice his belief that consumers should derive positive pleasure from the products they own. Besides Kodak, Teague has worked for other American blue-chip corporations: Ford, United States Steel, NCR, Du Pont, Westinghouse and Texaco. For Texaco he designed a corporate identity to be used on service stations which is still employed across the world. Teague also designed the Marmon 16 automobile in 1930 and was on the Board of Design at the 1939 New York World's Fair. He developed an aesthetic which held that beauty, which he called 'visible rightness', was inherent in all objects. Latterly he worked for Boeing, designing aircraft interiors and acting as a consultant for certain exterior visual details.

Bibliography

Walter Dorwin Teague *Design This Day – The Technique of Order in the Machine Age* Harcourt, Brace & Co, New York, 1940.

Wilhelm Wagenfeld

Kenneth Reid 'Walter Dorwin Teague, Master of Design', *Pencil Points* vol 18, September 1937, pp 539-570.
Walter Dorwin Teague 'How a Big Design Office Works', *Industrial Design* vol 2, no 1, pp 26-37.
[Not related to design, Teague also wrote *Land of Plenty and Possibilities* (1947) and *Flour for Man's Bread, a History of Milling* (1952)]

Photo Design Council

Craftsman and educator
b Bremen, Germany, 1900

Wagenfeld studied at the Weimar Bauhaus, and joined the Werkbund in 1928. He carried with him to the various establishments he worked in the influence of Moholy's teaching at the Metal Workshop there. From 1931 to 1935 he was at the Staatliche Kunsthochschule in Berlin and from 1935 to 1947 at the Lausitzer Glassworks in Weisswasser/Oberlausitz. Then he became professor at the Hochschule für Bildende Künste in Berlin and the Referent für Industrielle Formgebung in Stuttgart. Since 1954 he has run his own studio in Stuttgart and many of his designs for lamps and other metal products have won places in museums as solid examples of a certain sort of *modern* design.

Bibliography
Wilhelm Wagenfeld *Wesen und Gestalt der Dinge um Uns* Edouard Stichnote, Potsdam, 1948.
Zürich, Kunstgewerbemuseum *Industrieware von W Wagenfeld. Künstlerische Mitarbeit in der Industrie 1930-1960* 1960.
Paris, Centre de Création Industrielle *Du Bauhaus à l'Industrie: Wilhelm Wagenfeld, Objets Quotidiens* 1975.

Photo Frau Wagenfeld

Marco Zanuso

Architect and industrial designer
b Milan, Italy, 1916

Marco Zanuso is one of the most prominent living Italian industrial designers, well known for his transparent television sets for Brion-Vega and for his plastics furniture. As an architect, he designed the Olivetti plant in Buenos Aires in 1954. Zanuso took part in the 7th, 8th, 10th, 11th and 13th Triennali in Milan and has been a member of CIAM since 1956. Under Eliot Noyes he was employed for IBM and has been a member of the Commissione Edilizia del Comune di Milano (1961-1963). He teaches at the Faculty of Architecture in Milan Polytechnic.

Bibliography
Edilizia Moderna no 85, 1964, pp 23-25

Photo Design Council

DATES

1900 Exposition Universelle in Paris establishes Art Nouveau as a modern style.

1901 Marconi transmits radio across the Atlantic from Cornwall to Newfoundland.

1907 Foundation of the Deutscher Werkbund in Munich.

1915 Foundation of the Design and Industries Association in London.

1917 The word 'streamlining' begins to assume popular currency.

1919 Foundation of the Bauhaus in Weimar.

1919 Citroën Type A motor car appears.

1919 Raymond Loewy arrives in USA.

1920 First regular radio broadcasts from station KDKA, Pittsburgh, Pennsylvania, USA.

1921 Max Braun, a mechanic and engineer from East Prussia, starts a small factory at Frankfurt-am-Main. Its first product was a transmission belt connector. The factory began to assume a consciousness about design when associations were made with the Neue Frankfurt group.

1922 First edition of the Deutscher Werkbund journal, *Die Form*.

1924 Baird's mechanical-scan television system demonstrated.

1926 The Bauhaus moves to Dessau.

1927 Norman Bel Geddes opens his studio.

1927 Harley Earl designs his first motor car for General Motors, the La Salle V8.

1928 Raymond Loewy opens his studio.

1929 Henry Dreyfuss opens his studio.

1933 The Bauhaus, which, under political pressure, has moved to Berlin, closes down.

1934 Council for Art and Industry established in London.

1935 Braun builds the first battery-operated portable radio.

1935 IBM markets the first commercially successful electric typewriter.

1936 Alexandra Palace, London broadcasts the first television service.

1937 Harley Earl becomes President of General Motors Styling Division.

1937 The New Bauhaus opens in Chicago with László Moholy-Nagy as Director. Soon closes down, becoming the Institute of Design with the industrialist Walter Paepcke as patron.

1938 Ferdinand Porsche's Volkswagen and Pierre Boulanger's Citroën 2CV both running in prototype form.

1938 Braun develops the foil type electric razor.

1939 Chicago Futurama exhibition.

1940 Publication of important design books by Walter Dorwin Teague and Harold Van Doren.

1944 Council of Industrial Design established in London.

1947 Gordon Russell becomes Director of the Council of Industrial Design.

1949 First edition of *Design* magazine.

1951 Festival of Britain exhibition.

1953 First courses held at the Hochschule für Gestaltung, Ulm.

1953 First edition of *Stile Industria* under Alberto Rosselli.

1953 Raymond Loewy's Studebaker Starline coupé appears.

1954 First edition of *Industrial Design* magazine.

1955 The Compasso d'Oro award for good design established in Italy.

1955 First flight of Boeing 707 jetliner.

1956 The Independent Group organises 'This is Tomorrow' exhibition at the Institute of Contemporary Arts in London. This marks the beginning of the celebration of popular culture.

1958 Museum of Modern Art in New York organises a permanent display of Braun equipment.

1958 World's Fair in Brussels.

1960 Paul Reilly becomes Director of the Design Council in London.

BIBLIOGRAPHY

This bibliography is very selective and makes no attempt whatsoever to be complete. Two things in particular militate against any potential claim to comprehensiveness: the first is that almost all of what, say, Edgar Kaufmann and Reyner Banham have written is related in one way or another to the question of design, but it would be impractical to attempt a complete bibliography of the two; and the second is that much writing on design is repetitive, with ideas taken from one source reproduced willy-nilly without acknowledgment.

What this bibliography intends to do is make a section through all the body of writings about modern industrial design and present one version of a history. Although it is a partial view, I believe that it offers a fundamental understanding of the ideas which have concerned industrial designers and writers about design during this century.

Generally, the citing of periodical material has been restrained except where articles are of exceptional interest. The place of publication of all books is London, unless otherwise indicated.

An asterisk (★) indicates those publications from which texts have been quoted in the section beginning on page 25.

Ambasz, Emilio (ed) *Italy: the New Domestic Landscape–Achievement and Problems of Italian Design* New York, Museum of Modern Art and Florence, Centro Di, 1972. A luxuriously produced, well illustrated account of modern Italian design.

Architectural Review Whole issue on 'British Industrial Art', July 1933. A special edition of the *AR*, produced at the time when the magazine's interests were beginning to shift towards modernism.

Ashford, Fred C *Designing for Industry: Some Aspects of the Product Designer's Work* Pitman, 1956. A practical guide to product design by one of the few leading specialists.

Auer, Michel (trans D B Tubbs) *The Illustrated History of the Camera from 1839 to the Present* Kings Langley, The Fountain Press, 1975. A diverting, if slightly vacuous, picture book.

Banham, Reyner 'Design by Choice', *Architectural Review*, July 1961, pp 43-48. A survey of contemporary British industrial design.

'Household Godjets', reprinted from *New Society* in Paul Barker (ed) *Arts in Society* Fontana, 1977, pp 165-170. An article about domestic appliances which concentrates on those produced by the Braun firm.

'Servants of the Public Will', *Zeitschrift der Hochschule für Gestaltung* December 1965, pp 2-7. An article about the so-called 'New Bauhaus' at Ulm.

Theory and Design in the First Machine Age Architectural Press, 1960. The standard work on modern architecture and design, now in a paperback edition.

Barr, Alfred H *Art of the Machine* New York, Museum of Modern Art, 1934. A catalogue of an important exhibition of the 1930s.

★ Behrens, Peter 'Stil?', *Die Form* Heft 1, 1922. An essay on art and design by one of the first architect-designers to be retained by a large manufacturing organisation.

★ Bel Geddes, Norman *Horizons* John Lane, The Bodley Head, 1934 (English edition). The personal statement of a visionary designer; a classic text in its field.

Magic Motorways New York, Random House, 1940. The picture book which Geddes prepared in connection with the 1939 Motorama exhibition.

★ Bertram, Anthony *Design* Penguin Books, 1938. The published version of Bertram's influential radio talks of 1937.

The Enemies of Design DIA, 1946. A post-war recapitulation by Bertram; soon, however, his modernist views were to change.

Bill, Max *Die Gute Form* Winterthur, Buchdrückerei Winterthur AG, 1957. A small book about good design produced by one of the protagonists of the Ulm Hochschule für Gestaltung.

Form . . . a Balance Sheet of Mid-twentieth Century Trends in Design Basel, Verlag Karl Werner, 1952. A survey book by the architect of the Hochschule für Gestaltung, Ulm.

Bøe, Alf *From Gothic Revival to Functional Form. A Study in Victorian Theories of Design* Oslo University Press and Blackwell's, Oxford, 1957. Bøe's book is unjustifiably little known; it is one of the best studies of its subject.

Norsk/Norwegian Industrial Design Oslo, Kunstindustrimuseet, 1963. An illustrated survey.

Boumphrey, Halliday and Gloag 'What's Wrong with Design Today?', *The Listener* 19 April 1933, pp 607-610. A characteristically polemic piece from the period when *The Listener* was advocating modern design.

Braun-Feldweg, Wilhelm *Die Gute Industrieform, Zusammengestellt von der Neuen Sammlung* Munich, Die Neue Sammlung, 1955. Exhibition catalogue.

Normen und Formen Industrieller Produktion Ravensburg, Otto Maier Verlag, 1954.

Bush, Donald J *The Streamlined Decade* New York, Braziller, 1975. A well illustrated history of 1930s design in the USA.

Byggekunst Whole issue on industrial design, Oslo, no 2, 1961.

Carrington, Noel *Design* John Lane, 1935. Carrington worked mostly in publishing and was responsible for turning many people's tastes towards modern design; this was his first important book.

Industrial Design in Britain George Allen & Unwin, 1976. A retrospective view.

Celant, Germano *Marcello Nizzoli* Milan, Edizioni Comunita, 1968.

Cheney, Sheldon *Art of the Machine* New York, Whittelesey House, 1937. Inspired by the Museum of Modern Art Exhibition.

Chew, V K *Talking Machines 1877-1914 — Some Aspects of the Early History of the Gramophone* HMSO, 1967.

Constantin, Paul *Industrial Design* Curente Si Sinteze 9, Bucharest, Editura Meridiane, 1973. A very thorough survey of the whole profession of industrial design which includes a valuable discussion of the practice in Eastern Europe.

Coulson, Anthony J *A Bibliography of Design in Britain 1851-1970* Design Council, 1979.

★ Council for Art and Industry Report on *Design and the Designer in Industry* HMSO, 1937. An official reaction to popular agitation about encouraging good design.

Doblin, Jay *One Hundred Great Product Designs* New York, Van Nostrand Reinhold, 1970. The idea for this book, which originated at the Illinois Institute of Technology in 1957, is similar to the idea of this one. Doblin canvassed his colleagues to produce a list of 'great' product designs so that an IIT design course could have some criteria for evaluation.

Davis, Alec 'Popular Art Organised; the Manner and Methods of Raymond Loewy Associates', *Architectural Review*, pp 319-325, November 1951. The only article about perhaps the most influential of all American industrial designers.

de la Valette, John *Conquest of Ugliness. A Collection of Contemporary Views on the Place of Art in Industry* Methuen, 1935.

Dorfles, Gillo *Introduction à l'Industrial Design* Tournai, Casterman, 1974.

Il Disegno Industriale e la sua Estetica Bologna, Capelli Editore, 1963.

★Dreyfuss, Henry *Designing for People* New York, Simon & Schuster, 1955.

'The Industrial Designer and the Businessman' in *Harvard Business Review* November 1950, pp 77-85.

Eames, Charles See *Architectural Design* (whole issue) vol 36, September 1966.

Farr, Michael *Design in British Industry – A Mid Century Survey* Cambridge, The University Press, 1953. Effectively, a second edition of Pevsner's *Enquiry* of 1937.

Fischer Fine Art *Josef Hoffmann 1870-1956, Architect and Designer* 1977. Catalogue of a small private exhibition about the work of this designer from the Wiener Werkstätte.

Forty, Adrian 'Wireless Style: Symbolic Design and the English Radio Cabinet, 1928-1933', *Architectural Association Quarterly* vol 4, no 2, pp 23-31, 1972. The first modern study of its subject.

Fossati, Paolo *Il Design in Italia 1945-1972* Turin, Einaudi, 1972. A study of some leading Italian industrial designers. Well illustrated.

Freeman, John Wheelock 'The Studebaker Story', *Industrial Design* vol 1, no 1, February 1954, pp 38-45.

Gay, Bernard *Classics of Modern Design* exhibition catalogue, Camden Arts Centre, 1977. Well illustrated, but fails to make a distinction between industrial design and craft.

Giedion, Siegfried *Mechanization Takes Command – a Contribution to Anonymous History* New York, Oxford University Press, 1948. An immensely erudite and influential book about machines and culture across the centuries.

Gloag, John *Design in Modern Life* Allen & Unwin, 1934. Another of Unwin's design books; a typical text of the 1930s.

Industrial Art Explained Allen & Unwin, 1934.

Graeff, Werner 'Zur Form der Automobile', *Die Form* Jahrgang 6, 1931. An article by an author associated with the Deutscher Werkbund.

Grey, J *Fitness – for what Purpose?* DIA, 1946. One of the DIA booklets.

Hamburg *Moderne Deutsche Industrieform* Bildheft V, Museum für Kunst und Gewerbe, 1962. An illustrated survey.

★Hamilton, Richard 'Persuading Image', *Design* no 134 February 1960, pp 28-32.

Holme, C G *Industrial Design and the Future* Studio, 1934. This book has a real period flavour, but it is confusing in other respects because a proper distinction between craft-work and mass production is not made.

Industriekultur – Peter Behrens und die AEG 1907-1914 exhibition catalogue, Berlin, IDZ, 1978.

Irvin, Howard S 'The History of the Airflow Car', *Scientific American* August 1977, pp 98-106.

Kahn, Ely Jacques *Design in Art and Industry* New York, C Scribner's Sons, 1935.

Kaufmann, Edgar 'Borax, or the Chromium-Plated Calf' *Architectural Review* August 1948, pp 88-93. A criticism of vulgar, consumerist design by a protagonist of modernism.

'Design Sans Peur et Sans Ressources', *Architectural Forum* vol 125, no 2, September 1966, pp 68-70.

'Modern Design does not need Ornament', *College Art Journal* vol 6, Winter 1946, pp 140-142.

★ 'The Design Shift 1950-1960', *Industrial Design* vol 7, August 1960, pp 50-51.

'What is Modern Industrial Design?' *Museum of Modern Art Bulletin* vol 14, Fall 1946, pp 3-13.

Le Corbusier (pseud Charles-Edouard Jeanneret) *Aircraft* Studio, 1936. A picture book with a short introduction by the architect who did more than anybody else to introduce the machine as an influence in the visual arts.

Loewy, Raymond *The Locomotive* Studio, 1937. Loewy's contribution to the same series as Le Corbusier's *Aircraft*; other titles were planned, but none appeared.

Never Leave Well Enough Alone New York, Simon & Schuster, 1951. Loewy's witty autobiography. French edition: *La Laideur se Vend Mal* Paris, Gallimard. *L'Air du Temps* no 8, 1963, contains some additional material.

'Streamlined Transport', *Industrial Arts* vol 1, Autumn 1936, pp 174-182.

Industrial Design New York, The Overlook Press, 1979.

★Lux, Joseph-August *Ingenieur-Aesthetik* Munich, Verlag von Gustav Lammers, 1910. An argument in favour of the example of the machine in design by a leading writer of the Deutscher Werkbund.

MacCarthy, Fiona *All Things Bright and Beautiful* Allen & Unwin, 1972. Allen & Unwin continued its tradition of publishing books on design with this very useful work.

A History of British Design 1830-1970 Allen & Unwin, 1979. Revised, popular version of *All Things Bright and Beautiful*.

Mang, Karl *Geschichte der Modernen Möbel* Stuttgart, Gerd Hatje Verlag, 1978.

Meadmore, Clement *The Modern Chair – Classics in Production* Studio Vista, 1974. The best picture book on chairs available.

Milan *Esposizione Triennale Internazionale delle Arte Decorative e Industriali Moderne e dell' Architectura Moderna* 1954.

Forme Nuove in Italia Rome, Carlo Bestetti, 1957. A survey of recent Italian industrial design.

Mitarichi, Jane Fiske 'Harley Earl and his Product: the Styling Section', *Industrial Design* vol 2, no 5, pp 55-60.

Munich, Die Neue Sammlung, Staatliches Museum für Angewandte Kunst *Zwischen Kunst und Industrie. Der Deutsche Werkbund* exhibition catalogue, 1975. A well illustrated anthology.

Stuck Villa *Objekte der 20er Jahre* 1973-74 exhibition catalogue.

New York, Museum of Modern Art *The Design Collection – Selected Objects* 1970. No text, but many good pictures.

Noblet, Jocelyn de *Design – Introduction à l'Histoire de l'Evolution des Formes Industrielles de 1820 à Aujourd'hui* Paris, Stock-Chêne, 1974.

Olivetti *Olivetti* 1958. A well illustrated study of the firm, privately printed.

Paris, Centre de Création Industrielle, François Burkhardt et al (ed) *Ettore Sottsass Jr – de l'Objet Fini à la Fin de l'Objet* Paris, 1976. A small, illustrated study of one of Italy's most prominent living designers.

Paris, Musée des Arts Décoratifs *Formes Utiles* exhibition catalogue, 1949.
Formes Industrielles: Première Exposition Internationale exhibition catalogue, 1963.
Olivetti: Formes et Recherches exhibition catalogue, 1969.

★Pevsner, Nikolaus *An Enquiry into Industrial Art in England* Cambridge, The University Press, 1937.
'Broadcasting Comes of Age. The Radio Cabinet 1919-1940', *Architectural Review* May 1940, pp 189-190. An original article about one of the first domestic appliances to develop a style of its own.
Visual Pleasures from Everyday Things. An Attempt to Establish Criteria by which the Aesthetic Qualities of Design can be Judged Council for Visual Education, booklet no 4, Batsford, 1946.

Pick, Frank 'Meaning and Purpose of Design', *The Listener* 28 June 1933, pp 1016-1018. An article by the man who, by his interest in good, modern design, changed the appearance of London Transport.

Preussischer Kulturbesitz *Tendenzen der Zwanziger Jahre* exhibition catalogue, Berlin, Dietrich Reimer Verlag, 1977. The catalogue of a gigantic exhibition on the art of the 1920s; the section on 'Funktionelle Gestaltung' contains some good photographs.

Rateilli, Enzo 'Fortuna e Crisi del Design Italiano', *Zodiac* vol 20, 1970, pp 116-127. A critical look at Italian design.

Read, Herbert *Art and Industry* Faber, 1934. A new edition, with extra plates, was produced under the title *Kunst und Industrie* by Gerd Hatje Verlag, Stuttgart, nd. Read's *Art and Industry* is one of the classic texts on modern design.
The Future of Industrial Design DIA, 1946. One of the DIA booklets.

Redmayne, Paul *The Changing Shape of Things* John Murray, 1945.

Reid, Kenneth 'Masters of Design 2: Norman Bel Geddes', *Pencil Points* vol 18, January 1937, pp 2-32. The only illustrated biography of the designer.
'Master of Design, Walter Dorwin Teague', *Pencil Points* vol 18, September 1937, pp 539-570. The only biography of Teague. These two were the only articles devoted to industrial designers in a series which concentrated on architects.

Richards, G Tilgham *The History and Development of Typewriters* HMSO, 1964, 2nd edition, 1975. A dutiful history of typewriters.

Richards, J M 'Towards a Rational Aesthetic: an Examination of the Characteristics of Modern Design with Particular Reference to the Influence of the Machine', *Architectural Review* December 1935, pp 211-218. This article was J M Richards's contribution to the 1930s debate on modern design.

Rosselli, Alberto 'Disegno per l'Industria', *Domus* no 286, September 1953, pp 59-64. An example of the occasional series of articles on industrial design published by the Italian magazine.

Schmittel, Wolfgang *Design-Concept-Realisation* Zürich, ABC Edition, 1975. A beautifully produced study of Citroën, Braun, Sony and Swissair.

Selle, Gert *Ideologie und Utopie des Design. Zur Gesellschaften Theorie der Industriellen Formgebung* Cologne, DuMont Schauberg, 1968.

Smithson, Alison and Peter 'Concealment and Display: Meditations on Braun', *Architectural Design* vol 36, July 1966, pp 362-363. A rather vague article on the German firm.

★ Teague, Walter Dorwin *Design This Day – The Technique of Order in the Machine Age* New York, Harcourt, Brace & Co, 1940. Teague's own professional 'biography'.

★ Van Doren, Harold *Industrial Design – A Practical Guide* New York, McGraw-Hill Book Company, 1940.

Victoria & Albert Museum *The Wireless Show – 130 Classic Radio Receivers – 1920s to 1950s* exhibition catalogue, 1977. A brief, but informative, survey of radio cabinet design.

Waentig, Heinrich *Wirtschaft und Kunst. Eine Untersuchung über Geschichte und Theorie der Modernen Kunstgewerbebewegung* Jena, Gustav Fischer, 1909.

Zorzi, Renzo 'Olivetti – Concept and Form', *Graphis* no 150, 1971-1972, pp 346-381.

Zürich, Kunstgewerbe Museum *Die 20er Jahre: Kontraste eines Jahrzehnts* exhibition catalogue, 1973. A retrospective look at modern design.

Industrieware von W Wagenfeld. Künstlerische Mitarbeit in der Industrie, 1930-1960 nd. An exhibition catalogue about the metalwork of Wilhelm Wagenfeld, author of *Wesen und Gestalt der Dinge um Uns,* a consideration of the influence of the Bauhaus on the everyday world.

Modelfall Citroën – Produktgestaltung und Werbung exhibition catalogue, 1967. A beautifully produced celebration of Citroën, the one car manufacturer which pays most attention to design.

Stile Olivetti: Geschichte und Formen einer Italienischen Industrie exhibition catalogue, 1961. An interesting survey, but adds very little to the firm's own publication of 1958.

ACKNOWLEDGMENTS

It is easy to become fawning and generous with ritualised obsequiousness in writing acknowledgments, and because more people than I can hope to mention by name contributed to this book, I don't want to miss the opportunity to thank those who follow who are not anonymous. First of all Lord Reilly who, when he was still at the Design Council, encouraged me to write this book and introduced me to Paul Burall, head of the Publications Department there, who saw to it that I did. Nicola Hamilton edited the text and made enormous improvements while doing it. Gill Streater designed the book. The whole idea was enhanced by occasional conversations and frequent arguments with my friends, Quentin Hughes, Edward Cooper and Sanda Miller. Sarah Golding, Roger Cardinal and Agnès Cardinal all helped with the translations, while Sylvia Katz found all the photographs I wanted and plenty of others besides. From manufacturing industry Eberhard Fuchs of AEG-Telefunken, Ed Chamberlain of IBM (UK), David Maroni of Olivetti (GB), Dieter Schmitt of Braun and Joseph H Karshner of General Motors were all especially helpful. From public service and the world of collecting, I should like to isolate Ursula Prinz of the Neue-Nationalgalerie in Berlin, Wend Fischer of the Neue Sammlung in Munich, and Manfred Ludewig. Lastly, but she knows not least, I must thank Marion Johnson, who typed all of this with unfluctuating good humour.

INDEX

Italic numerals refer to photographs